...schriften

Institut für Systembiotechnologie
Universität des Saarlandes

Herausgegeben von Prof. Dr. Christoph Wittmann

Band 3

Cuvillier-Verlag
Göttingen, Deutschland

Herausgeber
Univ.-Prof. Dr. Christoph Wittmann
Institut für Systembiotechnologie
Universität des Saarlandes
Campus A1.5, 66123 Saarbrücken
www.iSBio.de

Hinweis: Obgleich alle Anstrengungen unternommen wurden, um richtige und aktuelle Angaben in diesem Werk zum Ausdruck zu bringen, übernehmen weder der Herausgeber, noch der Autor oder andere an der Arbeit beteiligten Personen eine Verantwortung für fehlerhafte Angaben oder deren Folgen. Eventuelle Berichtigungen können erst in der nächsten Auflage berücksichtigt werden.

Bibliographische Informationen der Deutschen Nationalbibliothek
Die Deutsche Nationalbibliothek verzeichnet diese Publikation in der Deutschen Nationalbibliographie; detaillierte bibliographische Daten sind im Internet über *http://dnb.d-nb.de* abrufbar.
1. Aufl. – Göttingen: Cuvillier, 2019

© Cuvillier-Verlag · Göttingen 2019
  Nonnenstieg 8, 37075 Göttingen
  Telefon: 0551-54724-0
  Telefax: 0551-54724-21
  www.cuvillier.de

1. Auflage, 2019
Gedruckt auf umweltfreundlichem, säurefreiem Papier aus nachhaltiger Forstwirtschaft.

ISBN 978-3-7369-9948-0
eISBN 978-3-7369-8948-1
ISSN 2199-7756

Whole plant *in vivo* and *in silico*

metabolic flux analysis:

towards biotechnological application

Dissertation

zur Erlangung des Grades

der Doktorin der Naturwissenschaften

der Naturwissenschaftlich-Technischen Fakultät III

Chemie, Pharmazie, Bio- und Werkstoffwissenschaften

der Universität des Saarlandes, Saarbrücken

von

M.Sc. Veronique Beckers

aus Antwerpen, Belgien

Saarbrücken

2. November 2015

Tag des Kolloquiums: 10. Februar 2016

Dekan: Professor Dr.-Ing. Bähre

Berichterstatter: Professor Dr. Christoph Wittmann

Professor Dr.-Ing Elmar Heinzle

Vorsitz: Professor Dr. Gert-Wieland Kohring

Akad. Mitarbeiter: Dr. Björn Becker

Vorveröffentlichungen der Dissertation

Teilergebnisse aus dieser Arbeit wurden mit Genehmigung der Naturwissenschaftlich-Technische Fakultät III, vertreten durch den Mentor der Arbeit, in folgenden Beiträgen vorab veröffentlicht:

Publications

Dersch, L.*, Beckers, V.* and Wittmann, C. (2016b) Green pathways: Metabolic network analysis of plant systems, *Metabolic engineering*, 34, 1-24. doi: 10.1016/j.ymben.2015.12.001

Dersch, L., Beckers, V., Rasch, D., Melzer, G., Bolten, C.J., Kiep, K., Becker, H., Bläsing, O.E., Fuchs, R., Ehrhardt, T. and Wittmann, C. (2016a) *In vivo* Assimilation, Translocation and Molecular Carbon and Nitrogen Fluxes in the Crop *Oryza sativa*, *Plant physiology*, accepted

* Both authors contributed equally to this work

Publications currently under peer-review

Beckers, V., Dersch, L., Lotz, K., Melzer, G., Bläsing, O.E., Fuchs, R., Ehrhardt, T. and Wittmann, C. (2016) *In silico* metabolic network analysis of plant metabolism, *BMC Systems Biology*, submitted

Beckers, V.*, Dersch, L.*, Rasch, D., Bläsing, O.E., Chen, X., Völler, G., Lotz, K., Fuchs, R., Ehrhardt, T. and Wittmann, C. (2016) Elucidating herbicide effect on rice crop metabolism by isotopically non-stationary ^{13}C metabolic flux analysis, *to be submitted*

* Both authors contributed equally to this work

Other publications

Jazmin, L.J., Beckers, V. and Young, J.D. (2014) User manual for INCA, published in Young, J. D. (2014). INCA: a computational platform for isotopically non-stationary metabolic flux analysis. *Bioinformatics, 30*(9), 1333-1335

Conference contributions

Melzer, G., Driouch, H., Beckers, V. and Wittmann, C., Systems-level design of filamentous fungi- integration of *in silico* flux modes and *in vivo* pathway fluxes towards desired production properties, 12th international conference on systems biology, August 28th – September 1st, 2011, Heidelberg/Mannheim, Germany

Patent applications

Fuchs, R., Woiwode, O., Peter, E., Schön, H., Wittmann, C., Rasch, D., <u>Beckers, V.</u> and Dersch, L. (2014) Method and device for marking isotopes, WO2014/079696 A1

Fuchs, R., Lotz,K., Leps, M., Jolkver., E, Wittmann, C. and <u>Beckers, V.</u> (2016) Elementary mode analysis for improving plant traits, priority application submitted on 12.02.2015, EP15154847

Danksagung

Zuerst möchte ich mich ganz herzlich bei Prof. Dr. Christoph Wittmann bedanken, der als Doktorvater und Mentor meine wissenschaftliche Weiterbildung vorangetrieben hat. Seine anhaltende Unterstützung und Ermutigung bei allen möglichen Herausforderungen werde ich für immer in Erinnerung behalten.

Herrn Prof. Dipl.-Ing. Dr. Elmar Heinzle möchte ich meinen Dank dafür aussprechen, dass er die Begutachtung meiner Dissertation und die Aufgabe als Prüfer bei meiner Disputation übernommen hat.

Insbesonders möchte ich mich bei unserem 'Flux Team' für die hervorragende Zusammenarbeit bedanken. Wir waren ein sehr heterogenes, aber komplementäres Team mit den unterschiedlichsten Expertisen. Detlev Rasch als Guru der technischen Erfindung, Lisa Dersch als Königin der Isotopen-Analytik, Guido Melzer als mein Modellierpate und Christoph als Antrieb um alle in die richtige Richtung zu lenken. Zusammen haben wir eine Leistung erbracht, auf die wir stolz sein dürfen.

Ein besonderer Dank geht auch an Dr. Regine Fuchs für das Übernehmen der Projektleitung, als Kooperationspartnerin bei der Metanomics GmbH, deren ewiger Enthusiasmus immer wieder ansteckend war. Auch ihren Kollegen und Mitarbeitern bei der Metanomics GmbH möchte ich gerne für ihren Einsatz danken. Insbesondere danke ich Olaf Woiwode für seinen grünen Daumen, Marcel Schink für seine innovativen Ideen im Labor und Katrin Lotz, Dr. Oliver Bläsing und Xuewen 'Wilson' Chen für die zahlreichen Diskussionen und Gespräche in denen sie mich mit ihrer wissenschaftlichen Kompetenz unterstützt haben. Bei der Metanomics GmbH bedanke ich mich für die finanzielle Unterstützung meiner Arbeit.

Lisa Dersch danke ich für die tolle Zusammenarbeit, für ihre bereitschaft immer mal wieder mein Reflexionsbrett zu sein, für die gegenseitige Motivation beim Paper schreiben, für die Erleichterung der Berliner Laborzeit, sowie für das Verjagen des ein oder anderen totes Mädchens.

I would like to thank Prof. Jamey Young at Vanderbilt University in Nashville, TN, USA for welcoming me as a guest in his lab and providing us with his software platform, already in a very early stage.

Bei den Mitarbeitern des IBVT der TU Braunschweig und des iSBio-Teams der Universität des Saarlandes bedanke ich mich für die angenehme Zeit. Lisa, Anna, Judith und Jessica danke ich für die Freude auf der Arbeit.

Ganz ‚dolle viel' möchte ich mich auch bei meinem zukünftigen Mann, Sören Starck, bedanken, der u.a. als Emotionsstabilisator und Nachtschicht-Buddy in den letzten Monaten, einiges geleistet hat.

Zum Schluss möchte ich mich noch ganz herzlich bei meiner Familie bedanken:

Aan mijn lieve ouders. Zonder jullie zat ik niet waar ik nu zit. Zonder jullie steun was ik niet naar het buitenland vertrokken en zonder jullie constante aanmoediging en motiverende woorden had ik deze droom nooit kunnen waarmaken. Mama, papa, jullie hebben mij zowel letterlijk als figuurlijk door dik en dun gesteund, waarvoor ik jullie bedank. Mijn thesis draag ik dan ook op aan jullie, want zonder jullie is de wereld hol.

Table of Contents

Summary

In the present work, both *in silico* and *in vivo* methods for flux analysis in plants were successfully developed and applied for enhanced understanding of plant physiology. Taken together, the *in silico* metabolic simulations provide detailed molecular insights into plant functioning, particularly by linking *in vivo* with *in silico* data. The knowledge gained from such a systems-biological approach, together with the proposed high potential of plants as biotechnological production platforms, especially for compounds requiring much redox power, will help to establish plants as biotechnological factories.

For the first time, the *in vivo* metabolism of an agriculturally relevant crop, *O. sativa*, was investigated, through non-stationary ^{13}C-metabolic flux analysis. This allowed elucidation of the *in vivo* intracellular carbon partitioning in rice plants and of the plants' necessity for futile cycling of resources, thus, contributing significantly to our current knowledge on plant metabolism. In addition, the effect of imazapyr, an industrially relevant herbicide, on rice metabolism was inspected using the newly established workflow. This first real-life case-study provides a valuable proof-of-principle and enabled a deeper understanding of the immediate metabolic effects of the treatment. This method can now be adopted to other crops, cell lines and stress inducers, such as abiotic stresses, herbicides and fungicides, and therefore, has great potential in green biotechnology.

Zusammenfassung

In dieser Arbeit wurden *in silico* und *in vivo* Methoden für Stoffflussanalysen in Pflanzen erfolgreich entwickelt und angewendet, um das Verständnis der Pflanzenphysiologie zu verbessern. In Summe erlauben die metabolischen *in silico* Simulationen detaillierte molekulare Einblicke in die Funktionalität von Pflanzen, insbesondere durch die Verknüpfung von *in vivo* und *in silico* Daten. Das Wissen, das durch einen solchen systembiologischen Ansatz gewonnen wird, kann genutzt werden um Pflanzen als biotechnologische Produktionsplattformen zu etablieren.

Zum ersten Mal wurde der *in vivo* Metabolismus der Nutzpflanze *O. sativa* durch nicht-stationäre ^{13}C-metabolische Flussanalyse untersucht. Dies erlaubte die Aufklärung der intrazellulären Kohlenstoffverteilung von Reispflanzen und der Notwendigkeit von scheinbar nutzlosen Ressourcenkreisläufen. Diese Arbeit steuert daher wesentlich zum aktuellen Wissensstand des Pflanzenmetabolismus bei. Zusätzlich wurde der Effekt von Imazapyr, einem industriell relevanten Herbizid, auf den Metabolismus von Reis mit der neu etablierten Methode untersucht. Diese erste praxisnahe Studie stellt einen wertvollen Machbarkeitsbeweis bereit, und erlaubte darüber hinaus ein tieferes Verständnis der metabolischen Effekte bei der Behandlung. Diese Methode kann auch auf andere Pflanzen, Zelllinien und Stressauslöser, wie zum Beispiel abiotischem Stress, Herbizide und Fungizide, angewandt werden und hat damit ein großes Potential für die grüne Biotechnologie.

1. Introduction

1.1. Rationale

Increasing world population, shortage of arable land and the resulting growing demand for food, feed and raw materials are major drivers to create plant lines with increased performance, e.g. better resistance to disease and drought (Yan & Kerr, 2002). In addition, plants play a significant role in the developing bioeconomy (Dyer et al., 2008) and emerge as platforms for sustainable production of therapeutics, renewable chemicals and biofuels, purely from sunlight and carbon dioxide (Yan & Kerr, 2002). Such development of plants with specific compositional traits adds impetus to the general interest in enhanced crops (Rajasekaran & Kalaivani, 2013; Saha & Ramachandran, 2013).

Admittedly, the optimization of plant performance using genetic engineering techniques is still lacking systems-level understanding of the effect of genetic modifications (Cusido et al., 2014; Sweetlove et al., 2003). Knowledge-based metabolic engineering approaches are often hampered by the complexity and robustness of plant systems and our still limited understanding of their metabolism (Junker, 2014; Shachar-Hill, 2013). Profound knowledge appears ultimately important to guide metabolic engineers, which becomes immediately clear from the achieved success in breeding superior microorganisms. Meanwhile, industrial microorganisms are optimized on a global scale, through systems metabolic engineering (Ajikumar et al., 2010; Becker et al., 2011; Hwang et al., 2014; Kim et al., 2014a; Kind et al., 2014; Paddon et al., 2013; Poblete-Castro et al., 2013). Particularly, systems metabolic engineering has benefitted from knowledge on metabolic fluxes, i.e. *in vivo* activities of intracellular pathways and reactions, in providing targets for genetic improvement (Kelleher, 2001; Stephanopoulos, 1999). In this regard, the analysis of metabolic fluxes of plant systems promises a huge next step towards understanding of their metabolic functions and superimposed regulation mechanisms towards superior plant varieties.

In recent years, a powerful collection of flux modeling approaches has been developed to model and simulate stoichiometric metabolic networks of microorganisms. Some methods investigate physiological capabilities by *in silico* analysis of the underlying biochemical conversions, such as elementary flux mode analysis (Schuster et al., 1999; Terzer & Stelling, 2008) and extreme pathway analysis (Papin et al., 2002; Price et al., 2002), whereas others rely on experimental data to deliver necessary constraints, such as ^{13}C-metabolic flux analysis (^{13}C-MFA) (Sauer, 2006; Wittmann, 2007; Young et al., 2011), flux balance analysis (Grafahrend-Belau et al., 2009) and metabolic control analysis (Wang et al., 2004). To engage these methods for plant metabolic flux analysis, the compartmented nature and autotrophic lifestyle of plants, pose specific challenges. However, the emerging state-of-the-art methods and the accumulating information on plant genomes (Michael & Jackson, 2013) now enable systems metabolic engineering concepts to be extended to plant networks for the design of plants with improved performance.

1.2. Objective

The aim of the present work was the advancement of novel *in vivo* and *in silico* flux analysis strategies for future application in plant biotechnology. Hereby, the high degree of complexity and connectivity of plant metabolic networks was a central aspect to be considered. This was achieved by carefully assembling comprehensive genome-based plant networks for the two systems to be studied: *Arabidopsis thaliana* as a model plant and *Oryza sativa ssp. japonica* as agriculturally relevant crop. The *in silico* method of choice for tackling the complex physiology of *A. thaliana* leaves through flux, was elementary flux mode analysis, as it promised the most comprehensive analysis of metabolism. In addition to truly computational modeling of the created plant network, integrated analysis with experimental data should be evaluated for improved physiological understanding.

Additionally, a comprehensive workflow should be developed for *in vivo* ^{13}C-based metabolic flux analysis of entire plants, including an experimental setup, raw data analysis and flux estimation through parameter fitting. Here, the use of $^{13}CO_2$ as sole carbon substrate, posed specific challenges with regard to tracer application, analytical sensitivity and modeling under isotopically non-stationary conditions. Focus was put on modeling strategies, in particular, ^{13}C-isotopically instationary metabolic flux analysis. Here, the aim was unraveling the metabolic complexity of rice seedlings. The final objective was to apply the developed workflow to gain deeper insights into relevant case-studies, including the prominent herbicide, Imazapyr. Taken together, this work should highlight and expand the potential of plant metabolic modeling in future biotechnological applications.

2. Theoretical Background

2.1. *Arabidopsis thaliana* as important model plant

Although *Arabidopsis thaliana* (thale cress) is often considered a simple weed, this small flowering plant is, without doubt, the most thoroughly studied plant species available. Especially its short generation time, highly reduced genome, regeneration through self-pollination and small stature make *Arabidopsis* an excellent candidate for research in plant biology (Koornneef & Meinke, 2010). Consequently, in the pre-genomics era *A. thaliana* emerged as the pivotal plant species in many biological research fields including physiology, biochemistry, molecular biology and evolution. Furthermore, the early sequencing and annotation of its complete genome (Arabidopsis Genome Initiative, 2000) allowed *A. thaliana* to establish itself in disciplines such as functional genomics, systems biology and biotechnology (Van Norman & Benfey, 2009). Despite its agricultural irrelevance, it has become impossible to imagine plant research without *A. thaliana* and likely, this simple weed will remain a major contributor to future plant research (Fig. 2-1).

Figure 2-1: *Arabidopsis thaliana* **in plant research**

Left: Plant sampling of *Arabidopsis thaliana* rosette for biotechnological research. Right: Genetically modified *Arabidopsis thaliana* in the flowering stage. Both photos belong to BASF SE, Ludwigshafen, Germany.

2.2. Rice biotechnology - towards genetically superior transgenic rice

2.2.1. *Oryza sativa* as model crop

Oryza sativa (rice) is a model crop with both agricultural and economical relevance (Fig. 2-2).

Since the genome of rice was fully sequenced in 2005 (International Rice Genome Sequencing

Project, 2005), its appearance in functional genomics, proteomics, systems biology, green

biotechnology and crop improvement has rapidly emerged (Coudert et al., 2010; Itoh et al.,

2005; Jiang et al., 2012; Kim et al., 2014c; Xu et al., 2005; Yin & Struik, 2008). Pertinent

characteristics responsible for this growing interest in *O. sativa* as a model crop, include its large

significance to human nutrition (FAO, 2013), the relatively small genome size (Goff, 1999) and

straightforward transformation as compared to other grass species (Yan & Jiang, 2007). In

addition, there is a considerable genomic homology between rice and other cereal species,

which increases the likelihood that rice-specific behavior could lead to elucidation of orthologous

functions in other cereals (Goff, 1999).

Figure 2-2: *Oryza sativa* **in plant research**

Left: *Agrobacterium*-mediated genetic transformation of rice seeds in a petri dish. Right: *Oryza sativa* in the greenhouse. Both photos belong to BASF SE, Ludwigshafen, Germany.

2.2.2. Importance of rice in biotechnology

Due to urbanization, soil degradation and climate change, arable land is diminishing fast, whilst the demand for crops rises with the growing world population, diet shift and bio-fuel consumption. Consequently, there is an urgent need to improve crop yields, as we otherwise will not be able to feed the world in the near future. This is especially true for rice, as it is the most important food crop in the world (FAO, 2013). The classical Latin word for rice, 'oryza', and 'sativa', meaning cultivated, provided rice with its species name. *Oryza sativa* has been cultivated for more than 10 000 years and is a monocot angiosperm, i.e. a flowering plant with one seed-leaf (Molina et al., 2011; Sang & Ge, 2007). It further belongs to the family of *Poaceae*, or true grasses, and contains two major subspecies: *japonica* and *indica*. *Japonica* varieties are sticky, short-grained and usually cultivated in dry fields in upland Asia, whereas the *indica* varieties are typically non-sticky, long-grained and mainly lowland rice cultivars, grown mostly submerged throughout tropical Asia (Garris et al., 2005). Together, these two subspecies provide more than one third of the global population with their daily calories.

Technological advances during the green revolution enabled the distribution of pesticides, synthetic fertilizers, irrigation technologies and high-yield varieties acquired through conventional breeding techniques. This already led to a vast increase in productivity between 1960 and 2000. However, since then, yield has not improved significantly (Ray et al., 2013; Zhu et al., 2010). We have come to a time, where traditional breeding techniques have to be supplemented with knowledge from genome analysis, systems biology and plant biotechnology to realize a second green revolution through genetic engineering of crops (Sakamoto & Matsuoka, 2004). Genetic transformation of plants has come a long way since the first transgenic rice plant was generated about 25 years ago (Toriyama et al., 1988; Zhang et al., 1988; Zhang & Wu, 1988). Especially the development of reproducible genetic transformation protocols, either through direct DNA transfer or by *Agrobacterium*-mediated transformation technologies (Zhu et al., 2010), enabled the construction of genetically engineered rice varieties with improved characteristics.

Particularly usefull for current and future genetic manipulation towards desired plant traits is the recent progress on the front of genome editing, which permits site-specific changes to rice DNA through the use of designer nucleases (Li et al., 2015; Petolino, 2015; Yu et al., 2015).

Despite the substantial increase in crop protection during the green revolution, crop loss, caused by various biotic and abiotic factors, is still eminent. The introduction of agronomically useful genes in rice, could significantly reduce these losses. Examples include resistance to drought (Jeong et al., 2010; Joo et al., 2014; Qian et al., 2015), salinity (Campo et al., 2014; Ghosh et al., 2014; Sahoo et al., 2014), extreme temperatures (Li et al., 2013; Qin et al., 2015; Yang et al., 2013), oxidative stress (Kim et al., 2014b; Lee et al., 2013; Park et al., 2013), mineral deficiency (Takahashi, 2003), insect predation (Qi et al., 2009; Quilis et al., 2014), fungal infestation (Chujo et al., 2014; Qian et al., 2014), viral invasion (Ma et al., 2011; Sasaya et al., 2014; Shimizu et al., 2013) and bacterial infection (Goto et al., 2014; Lu et al., 2014). One prominent example is the introduction of endotoxin-producing genes from *Bacillus thuringiensis* into rice, as it offers protection against lepidopteran pests, which are responsible for 2–10 % of the total rice yield loss in Asia (High et al., 2004). Without a doubt, this list is not comprehensive and merely provides a glimpse on the variety of stress-resistance genes recently engineered into rice.

Furthermore, nutritional improvement of rice can help reduce malnutrition, as it is currently the staple food in most developing countries (Bajaj & Mohanty, 2005; Bhullar & Gruissem, 2013). The most prominent example hereof is the bio-fortification of rice with provitamin A, giving the grains a golden color, hence golden rice (Bhullar & Gruissem, 2013). Vitamin A deficiency is an important cause of eye-defects, leading to permanent blindness when untreated. In developing countries around 250 million pre-school children and a substantial proportion of pregnant women are estimated to suffer from severe vitamin A deficiency (WHO, 2015). Even prior to blindness, vitamin A-deficiency increases child mortality as a result of enhanced susceptibility to measles, diarrhea, and malaria (UNICEF, 2009). Other examples of nutritionally enriched rice varieties include bio-fortification with folate (Storozhenko et al., 2007), iron (Masuda et al., 2012; Wirth et al., 2009), zinc (Johnson et al., 2011), essential amino acids (Lee et al., 2001; Lee et al., 2003; Long et al., 2013; Wakasa et al., 2006) and improved oil quality (Anai et al., 2003). Additionally, transgenic rice can be a production platform for heterologous proteins that can be applied as edible vaccines (Suzuki et al., 2011; Yang et al., 2012; Zhang et al., 2009) and medicine (Xie et al., 2008).

Despite the many successes in developing transgenic rice with improved nutritional content or resistance to both biotic and abiotic stress, a ceiling in yield improvement has been reached. Therefore, additional strategies towards higher yields, such as maximizing the conversion efficiency of CO_2 and light into biomass, are desperately needed (de Bossoreille de Ribou et al., 2013). Recent efforts towards developing such transgenic lines have mainly focused on photosynthetic efficiency by improving transient pools of sink starch (Gibson et al., 2011; Smidansky et al., 2003), introducing the C_4 photosynthetic machinery and the cyanobacterial CO_2 concentrating mechanism into rice (Price et al., 2013; von Caemmerer et al., 2012), as well as manipulating single photosynthetic functions such as RuBisCO (Lin et al., 2014; Parry et al., 2013) and sedoheptulose 1,7-bisphosphatase (Zhu et al., 2010). However, there is limited knowledge about the carbon conversion efficiency in rice, or any other crop (Alonso et al., 2007; Goffman et al., 2005), so that redirection of metabolic carbon flow to achieve higher yields has not been successful yet (de Bossoreille de Ribou et al., 2013). Overall, it can be concluded that recent strategies to improve yield have been hampered by lacking knowledge on the systems-wide physiology of rice (Long, 2014; Yamamoto et al., 2009). Therefore, improving the systems-level understanding of plants will form a stepping stone towards the development of high-yield crops (de Bossoreille de Ribou et al., 2013).

2.3.   Plant metabolism

2.3.1. Spatial separation and temporal shift in carbon and energy acquisition

The primary site for assimilation of atmospheric CO_2 and energy-gain from light through photosynthesis, is the plant leaf. From here, photosynthetic assimilates, mainly sucrose and some amino acids, are transported to the roots and actively growing plant parts, where they function both as carbon and energy-source (Fischer et al., 1998; Lalonde et al., 2004). To fulfill their task with maximum efficiency, leaves are organized into specialized tissues and cell types, each playing a distinct physiological role (Fig. 2-3). These various cell-types exhibit a highly conserved intracellular organization. Like for all eukaryotes, their DNA is stored in the nucleus, which is surrounded by the cytosol. Next to the vacuole, the cytosol houses specialized organelles, such as microbodies, chloroplasts and mitochondria (Taiz & Zeiger, 2006). This high level of intracellular organization is also reflected in the allocation of specific metabolic functions to distinct organelles, e.g. chloroplasts and mitochondria are the energy-producing sites of the cell, responsible for photosynthesis and respiration, respectively (Araújo et al., 2014; Weber & Linka, 2011), whereas, peroxisomes, a type of microbody, neutralize highly toxic hydrogen peroxide and participate in photorespiration (Hu et al., 2012).

During the day, the assimilation of carbon fuels leaf metabolism and allows the accumulation of starch as well as the export of sucrose to non-photosynthetic tissues. Since plants have to resort to an alternative energy source at night, the stored transitory starch is degraded to maintain leaf metabolism in the dark. The dramatic metabolic shift from photoautotrophy during the day, to a heterotrophic life style at night in photosynthetic organs, is not apparent in heterotrophic plant tissues, as they are continuously provided with sucrose, delivering both carbon and energy to the cells. Both these temporal features and the previously discussed spatial organization should be considered in systems-wide plant investigations, as whole plant physiology is a tight cooperation between its organs, tissues, cells and organelles throughout the day-night cycle.

Organelle   Cell   Organ/Tissue   Whole plant

Chloroplast

Photosynthetic Cell   Leaf Tissue

Mitochondrion

Root Hair Cell   Root Apex

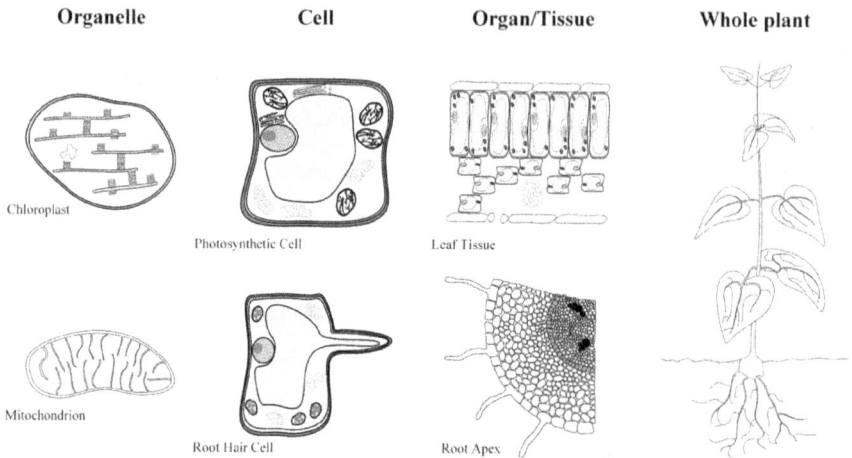

Figure 2-3: Organizational layers of plant anatomy

Whole plants are composed of different tissues, e.g. leaf tissue and root apex, which in turn consist of distinct cell types. Photosynthetic cells are characterized by large vacuoles and the presence of many chloroplasts and mitochondria, whereas root hair cells have 'hair-like' outgrowths and can, next to a vacuole and nucleus, contain amyloplasts and mitochondria. Specific metabolic functions are allocated to distinct organelles, such that whole plant physiology is a tight cooperation between its organs, tissues, cells and organelles throughout the day-night cycle. [1]

2.3.2. Cellular metabolism of photosynthetic C_3-leaf

At the heart of the cellular metabolism of photosynthetic leaf tissue is the Calvin–Benson–

Bassham (CBB) cycle, also known as reductive pentose phosphate cycle or C_3-cycle (Fig. 2-4).

The CBB cycle starts with the carboxylation of ribulose 1,5-bisphosphate by RuBisCO, yielding

two molecules 3-phosphoglycerate. These are subsequently reduced in a two-step process to

build glyceraldehyde 3-phosphate. The third phase of the cycle is responsible for the

regeneration of ribulose 5-phosphate from glyceraldehyde 3-phosphate through several

enzymatic steps, which are strongly connected to the non-oxidative pentose phosphate pathway.

Typically, one molecule 3-phosphoglycerate is gained for every three successions. As the CBB

---

[1] This figure and all components hereof are hand-drawn by the author of this work.

cycle is responsible for the energetically expensive fixation of CO_2, careful regulation is in place

to optimize energy use. Consequently, several stromal enzymes of the CBB cycle are switched

off during the dark, including: RuBisCO, fructose 1,6-bisphosphate phosphatase,

sedoheptulose 1,7-bisphosphate phosphatase, ribulose 5-phosphate kinase and NADP-

glyceraldehyde 3-phosphate dehydrogenase. This light-dependent modulation occurs both on

the transcriptional and post-translational level (Bläsing et al., 2005; Buchanan & Balmer, 2005;

Mora-Garcia et al., 2006).

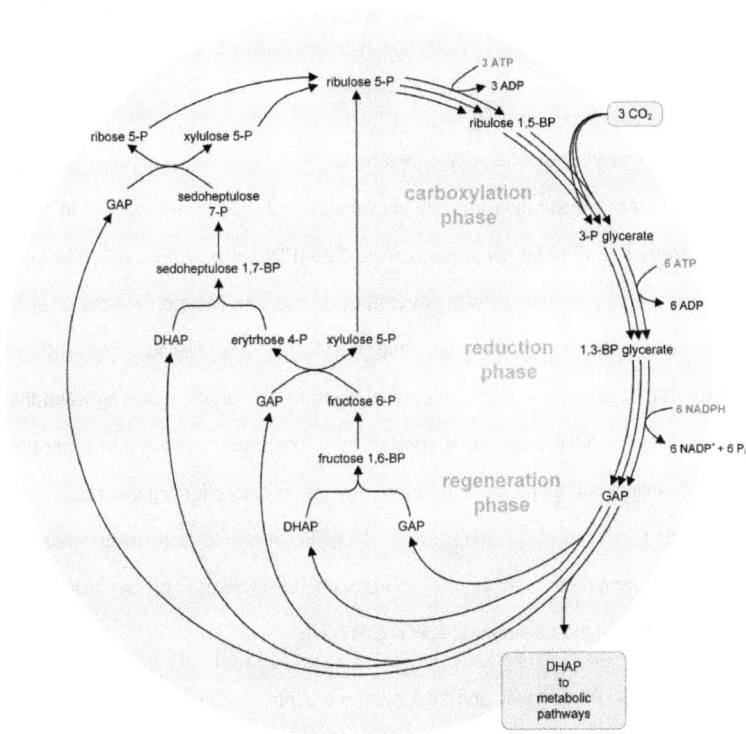

Figure 2-4: The Calvin-Benson-Bassham cycle, the heart of photoautotrophic carbon assimilation

Calvin-Benson-Bassham cycle consisting of three principle phases: carboxylation, reduction and regeneration. To gain one triose-phosphate, three molecules of carbon dioxide have to be assimilated by completing the cycle three times. Adjusted from Berg et al. (2007).

Additionally, the process of photorespiration is tightly coupled to the CBB cycle as RuBisCO is also capable of catalyzing the oxidation of ribulose 1,5-bisphosphate to 3-phosphoglycerate and 2-phosphoglycolate, which forms the first step in photorespiration (Fig. 2-5). Phosphoglycolate can be salvaged by conversion to 3-phosphoglycerate through several enzymatic steps that are distributed across the chloroplast, the peroxisome and the mitochondrion. Generally, two molecules of 2-phosphoglycolate are converted into one molecule of 3-phosphoglycerate and one molecule of CO_2. Consequently, photorespiration and CBB cycle operate in opposite directions, which is rather counterproductive. Moreover, CO_2 and O_2 compete at the active site of RuBisCO and both low CO_2 levels and high temperature promote oxygenation (Jensen, 2000).

An important node in photoautotrophic metabolism is 3-phosphoglycerate as it is not only the end-product of photorespiration and photosynthesis, but also forms the link to the Embden–Meyerhof–Parnas (EMP) and the pentose phosphate (PP) pathway (Fig. 2-5). The non-oxidative PP pathway is tightly intertwined with the CBB cycle and can occur only in the plastid, whereas the EMP and oxidative PP pathway have a cytosolic and plastidic copy (Kruger & von Schaewen, 2003). Specific transporters are in place to exchange carbon between the different organelles. The plastidial envelope is specialized in channeling sugars and sugar-phosphates, whereas the mitochondrial membrane has mainly translocators for organic acids (Facchinelli & Weber, 2011; Furumoto et al., 2011; Laloi, 1999). For instance, a pyruvate transporter in the mitochondrial membrane enables the end-product of glycolysis to be channeled into the mitochondrion to fuel the tricarboxylic acid (TCA) cycle.

The EMP pathway, PP pathway and TCA cycle are central metabolic pathways as they provide precursors to many biosynthetic pathways in higher plants (Fig. 2-5). For instance, lipids and pigments are synthesized from acetyl-CoA, and the hexose-phosphate pool provides building blocks for cellulose synthesis, whereas nucleotides are derived from pentose phosphate, and phosphoenolpyruvate is incorporated in aromatic amino acids and lignin. However, the vast spectrum of secondary metabolites and the structural diversity in some polymers such as lignin and hemi-cellulose, impede detailed knowledge on their biosynthesis and regulation (Pauly et al., 2013). Consequently, many cellular pathways and their time-dependent interplay are still undiscovered or poorly understood.

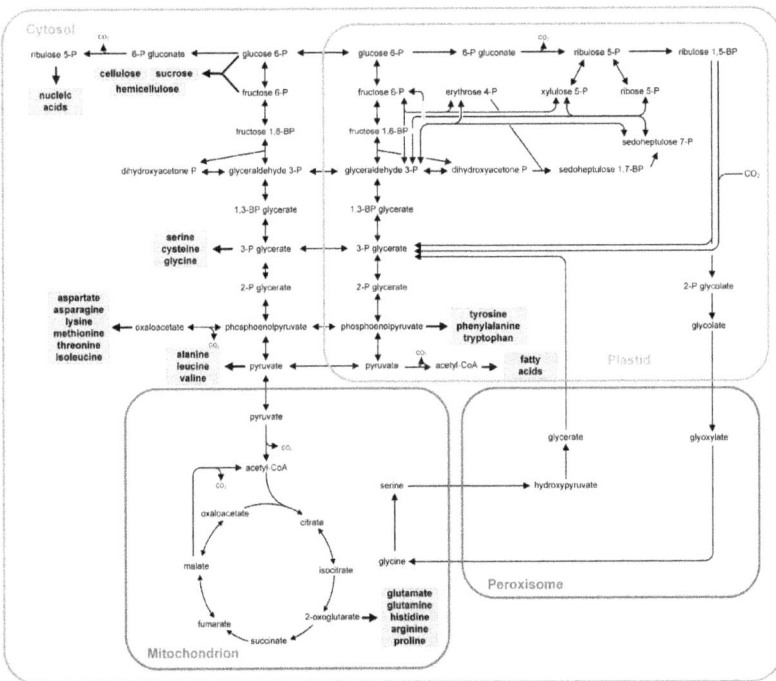

Figure 2-5: Metabolic pathways of central carbon metabolism

Compartmented view on central carbon metabolism of autotrophic plant leaf cell, including metabolic precursor demand for anabolism. Central carbon metabolism encompasses cytosolic and plastidic EMP and oxidative PP pathway, mitochondrial TCA cycle, plastidic CBB cycle and PP pathway as well as photorespiration.

2.4. Metabolic network modeling

The vast experience from past decades in breeding superior bacteria, clearly teaches that successful engineering has to build on understanding (Becker et al., 2011; Hwang et al., 2014; Kim et al., 2014a; Kind et al., 2014; Poblete-Castro et al., 2013; Trinh, 2012). Particularly, metabolic engineering has benefitted from knowledge on metabolic fluxes, i.e. *in vivo* activities of intracellular pathways and reactions, in providing targets for genetic improvement (Kelleher, 2001; Stephanopoulos, 1999). Metabolic fluxes through biochemical networks arise from the catalytic action of proteins, encoded by transcripts, which interconvert the typically hundreds of metabolites for the various needs of the cell. Thus, fluxes are the integrated output of transcripts, proteins and metabolites and offer a systems-level analysis of an organism (Kohlstedt et al., 2010). In this regard, the analysis of metabolic fluxes in plant systems promises a huge leap forward towards understanding of their metabolic functions and superimposed regulation mechanisms (Shachar-Hill, 2013).

However, unlike transcript levels or metabolite concentrations, metabolic fluxes cannot be quantified directly. Instead, *in vivo* fluxes are inferred from experimental ^{13}C isotopic labeling patterns, frequently in a combined approach with stoichiometric (Kruger et al., 2012) or kinetic modeling (Yuan et al., 2008). Additionally, fluxes can be predicted *in silico* by mathematical models, solely relying on genomic information and physiological characteristics, such as light influx and CO_2 assimilation rate (Kauffman et al., 2003). Both *in vivo* and *in silico* approaches require metabolic network information, so that recent sequencing of plant genomes and the reconstruction of genome-scale networks can be seen as important steps towards plant metabolic flux investigation (Table 2-1).

Table 2-1: Sequenced plant genomes that led to the construction of genome-scale metabolic networks.

Plant	Genome Size	Model Size (reactions x metabolites)	Organ
Chlamydomonas reinhardtii (green algae)	111 Mb	1725 x 1862 [2]	Cell culture
	17 737 genes [1]	2190 x 1068 [3]	Cell culture
Arabidopsis thaliana (thale cress)	135 Mb	1406 x 1253 [5]	Heterotrophic tissue culture
	27 000 genes [4]	1567 x 1748 [6]	Individual model of leaf, stem and root
		1363 x 1078 [7]	Individual model of leaf, stem, root, flower, silique, ...
		9727 x unspecified [8]	Multi-tissue model of leaf, stem and root
Brassica napus (rapeseed)	1130 Mb [9]	671 x 666 [10]	Seed
Oryza sativa (rice)	430 Mb	3316 x 2986 [12]	Leaf
	50 000 genes [11]	1736 x 1484 [13,14]	Leaf
Hordeum vulgare (barley)	5 100 Mb	257 x 234 [16]	Seed
	26 159 genes [15]	702 (269) x 890 [17]	Multi-organ model (transport)
Sorghum bicolor (sorghum)	730 Mb	1755 x 1588 [19]	Leaf
	34 496 genes [18]		
Zea mays (maize)	2 400 Mb	1755 x 1588 [19]	Leaf
	39 656 genes [20]	1985 x 1825 [21]	Leaf
		8525 x 9153 [22]	Leaf
		2322 x 2635 [23]	Leaf
		2304 x 2636 [23]	Embryo
		2280 x 2636 [23]	Endosperm

References: [1] Merchant et al. (2007) [2] de Oliveira Dal'Molin et al. (2011) [3] Chang et al. (2011) [4] TAIR [5] Poolman et al. (2009) [6] de Oliveira Dal'Molin et al. (2010a) [7] Mintz-Oron et al. (2012) [8] de Oliveira Dal'Molin et al. (2015) [9] Chalhoub et al. (2014) [10] Hay et al. (2014) [11] Kawahara et al. (2013) [12] Liu et al. (2013) [13] Poolman et al. (2013) [14] Poolman et al. (2014) [15] International Barley Genome Sequencing et al. (2012) [16] Grafahrend-Belau et al. (2009) [17] Grafahrend-Belau et al. (2013) [18] Paterson et al. (2009) [19] de Oliveira Dal'Molin et al. (2010b) [20] Gramene [21] Saha et al. (2011) [22] Simons et al. (2014) [23] Seaver et al. (2015)

Five main strategies in analyzing metabolic fluxes can be distinguished of which some approaches are independent from labeling data, and others rely on tracer studies using radioactive or stable isotopes (Fig. 2-6). Label-independent methods range from purely *in silico* investigations, entirely relying on network stoichiometry, i.e. elementary flux mode analysis, to more experiment-based methods such as flux balance analysis and kinetic modeling, which integrate uptake and production rates or kinetic parameters i.e. maximal enzyme activity, kinetic constants or metabolic concentration profiles. Labeling-based approaches infer *in vivo* fluxes from isotope-labeling experiments, either in combination with kinetic modeling, i.e. kinetic flux profiling (Yuan et al., 2008) or with stoichiometric modeling, i.e. ^{13}C-metabolic flux analysis (Kruger et al., 2012; Young et al., 2008). Each approach bears individual advantages and disadvantages. *In silico* methods are sensitive to network quality, however provide cheap, powerful alternatives to experiment-based methods. On the other hand, tracer-based methods offer a reliable estimate of *in vivo* fluxes and are, despite their high experimental demand, commonly applied. However, their low throughput is seen as a major disadvantage (Junker, 2014). In the remainder of this chapter, elementary flux mode analysis and tracer-based approaches for flux estimation in plant tissues are more thoroughly discussed, because of their relevance to this work. However, as both flux balance analysis and kinetic modeling are out of the scope of this thesis, they will not be considered further. The reader is kindly referred to Dersch et al. (2016b) for additional material on the topic.

2.4.1. Assembly of metabolic network model

In silico investigation of metabolism builds on a relevant set of pathways and enzymatic conversions, which are defined, based on the genome annotation of the specific plant species and available biochemical data. Careful reconstruction of the metabolic network is the most important step in model-based flux determination as it forms the basis of subsequent analyses (de Oliveira Dal'Molin & Nielsen, 2013; Schallau & Junker, 2010). Depending on the purpose of the study, such networks range from concise sets of only a few reactions in a specific pathway

(Assmus, 2005; Rohwer & Botha, 2001; Schwender et al., 2004) to comprehensive genome-scale reconstructions encompassing several hundreds to thousands of individual metabolic conversions (de Oliveira Dal'Molin et al., 2010a, 2010b; Grafahrend-Belau et al., 2009; Poolman et al., 2009; Saha et al., 2011; Seaver et al., 2015; Williams et al., 2010). Important information for each reaction involves its reaction stoichiometry, including cofactor use, its thermodynamic properties within the cellular environment and its localization.

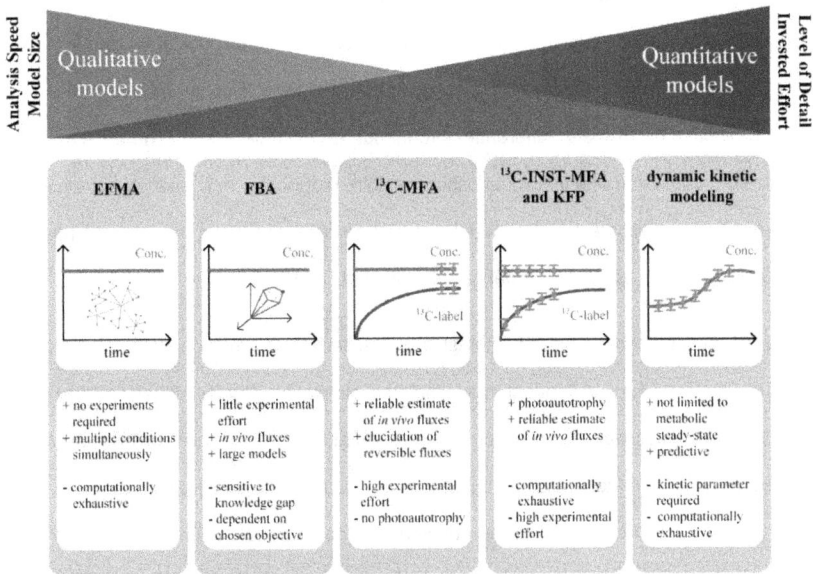

Figure 2-6: Methods for the analysis of fluxes

Overview on relevant flux analysis methods applicable to plants, with their respective advantages and disadvantages. A simple graphical representation of the conditions during modeling are given, i.e. concentration and labeling dynamics in red and blue, respectively. Typical sampling windows are indicated for experiment-based methods, whereas the applied *in silico* approach is depicted for strictly stoichiometric methods.

To which detail the selected reactions are assigned to a specific organ, tissue type or intracellular compartment, depends on the availability of accurate information specific to the organizational level. Network size and complexity are often constrained by the selected modeling approach, as too many reactions and metabolites might render the computational processing inefficient. Although compartmented models are more realistic, they are still strongly hampered by a lack of information on the correct allocation of every reaction and the existing transporters. Notwithstanding, the incorporation of compartmentation delivers high-quality descriptions of plant cellular metabolism (Pilalis et al., 2011; Poolman et al., 2013; Williams et al., 2010). Furthermore, whole plants consist of many different organs (leaf, stem, seeds, roots, flowers, etc.) that operate differently throughout the diurnal cycle. Therefore, multi-tissue metabolic models that link individual tissue models, require not only spatial, but also temporal separation to investigate systems-wide features such as growth or source-sink relations (de Oliveira Dal'Molin et al., 2015; Grafahrend-Belau et al., 2013).

Once the metabolic reactions of interest are selected, they are brought into a mathematical notation as basis for subsequent modeling and simulation studies. Most typically, they are expressed as mass balances for its individual metabolites in the form of ordinary differential equations (ODEs). Mass balance equations are defined for every internal metabolite, thus yielding a set of ODEs, which can be denoted in a stoichiometric matrix $S_{(m \cdot n)}$, with n reactions and m metabolites.

$$\frac{dS}{dt} = S \cdot v$$ (Eq. 1)

Here, v is the flux vector, containing all internal reaction rates of the metabolic network. In case, metabolic (pseudo)steady-state is assumed, ODEs are simplified into a set of algebraic equations:

$$\frac{dS}{dt} = S \cdot v = 0$$ (Eq. 2)

For plant systems, this has proven to be a valid assumption in exponentially growing cell cultures, mature differentiated tissues and developing embryos during seed filling (Libourel & Shachar-Hill, 2008). Many alternative flux distributions can obey these mathematical balances resulting in a flux space that is constrained by the network topology. The entire set of feasible fluxes can be analyzed to investigate the functional capabilities of the metabolic network through enumeration of elementary flux modes by mathematical software packages (Lotz et al., 2014; Terzer & Stelling, 2008; von Kamp & Schuster, 2006; Zanghellini et al., 2013). Other *in silico* approaches solve the underdetermined mass balancing problem by finding the optimal flux distribution for a particular objective, e.g. flux balance analysis (Lotz et al., 2014), or include an additional layer of information about the kinetic rate laws associated with the specific reactions (Rohwer, 2012; Schallau & Junker, 2010).

2.4.2. Elementary flux mode analysis

Feasible metabolic behavior and the boundaries of steady-state fluxes can be explored by structural analysis of the network topology. For instance, the computation of elementary flux modes (EFMs) yields all possible independent metabolic routes in steady-state, solely relying on genomic information (Schuster et al., 1999). Herein, every EFM represents a thermodynamically and stoichiometrically feasible pathway that is unique, minimal and can operate independently. The weighted sum of these minimal pathways describes the entire flux solution space (Schuster et al., 1999), and therewith the metabolic capabilities of the network. Mathematical algorithms to compute the EFMs are available (Klamt et al., 2005; Schuster et al., 1999; Terzer & Stelling, 2008; Wagner, 2004) and readily implemented in user-friendly tools such as METATOOL (Pfeiffer et al., 1999; von Kamp & Schuster, 2006) and efmtool (Terzer & Stelling, 2008). However, EFM computation is exhaustive and although increasingly faster methods, able to process increasingly larger networks, were developed (Acuña et al., 2009; Klamt et al., 2005; Rezola et al., 2011; Terzer & Stelling, 2008), structural pathway analysis is still confronted with the problem of scalability to genome-wide models (Zanghellini et al., 2013).

2.4.2.1. Applicability of elementary flux mode analysis in plant systems

EFM-based analysis has been a powerful tool in microbial rational engineering (Becker et al., 2011; Poblete-Castro et al., 2013; Trinh, 2012), with typical models comprising between 10 and 100 metabolic reactions. Although only few studies in plant systems exist (Beurton-Aimar et al., 2011; Rohwer & Botha, 2001; Schwender et al., 2004), they provide clever insights into their metabolic functioning. For instance, through EFM analysis, a previously undescribed metabolic route between carbohydrate and oil in *Brassica napus* seeds was revealed (Schwender et al., 2004). Additionally, a small metabolic model of *Brassica napus* seeds allowed comparison of the flux solution space under different environmental conditions by considering relative flux change and flux efficiency (Beurton-Aimar et al., 2011). The predicted flux changes showed very high consistency with internal flux rearrangement as determined by labeling-based *in vivo* flux estimation (Beurton-Aimar et al., 2011), indicating that *in silico* EFM analysis is a valid tool to gather valuable insights into seed metabolism.

At present, EFM models for plant systems merely provide a minimalistic description of organ-specific metabolism. Although pertinent characteristics of plant physiology, such as carbon conversion efficiency, were further elucidated with these models, today EFM analysis might not be suitable to address the complexity of systems-wide whole plant metabolism. The number of calculated elementary modes explodes combinatorially with increasing model complexity, rendering the integration of comprehensive models for different organs and tissue-types impossible, as mathematical methods and computational efficiency are at present not scalable to this size (Zanghellini et al., 2013). Additionally, it gets more difficult to gain biological insights from significantly larger EFM matrices.

2.4.2.2. Post-processing of modes: seeing the wood for the trees

The number of calculated elementary flux modes grows rapidly with increasing complexity and size of the network, from several hundreds to more than 70 million modes (Table 2-2). Since increased computational performance and more efficient algorithms now allow the generation of very large sets of modes, the current challenge is the analysis of this vast amount of flux data to gather valuable physiological insights. Therefore, the resulting EFM matrix is often reduced to a manageable set of fluxes by using separate models for specific conditions (Schäuble et al., 2011; Taffs et al., 2009) or by merely using a subset of the calculated elementary flux mode matrix (Melzer et al., 2009). For instance, a general network of central carbon metabolism for *Chlamydomonas reinhardtii* was extended with the biosynthesic reactions of five particular amino acids (Schäuble et al., 2011). In this study, the night-time metabolism of each amino acid was investigated by separate elementary flux mode analyses, unraveling the effect of experimentally observed downregulation of particular enzymes during the night. This study revealed valuable insights into circadian regulation, however, it remains questionable whether important physiological aspects were not lost due to such segregated analysis of metabolism.

Table 2-2: Size of flux space for five increasingly complex metabolic networks.

Number of elementary modes	Number of reactions	Number of metabolites	Organism	Source
289	62	59	*C. glutamicum*	Melzer et al. (2009)
16.084	121	106	*Aspergillus niger*	Melzer et al. (2009)
74.507	118	68	Bacterial consortium	Taffs et al. (2009)
1.352.352	105	95	*C. reinhardtii*	Schäuble et al. (2011)
71.266.960	230	218	*S. cerevisiae*	Jol et al. (2012)

One way to extract relevant information from elementary flux modes is by determining the maximum theoretical biomass or product yield ($Y_{P/C}$) as follows (Melzer et al., 2009):

$$\max Y_{P/C} = \max\left(\frac{s_{P,j}\ \xi_{P,j}}{s_{C,j}\ \xi_{C,j}}\right) \qquad 1 \leq j \leq n \qquad (\text{Eq. 3})$$

This value allows assessment of production capacity and growth optimality of an organism, as well as comparison with experimental yields to evaluate metabolic engineering potential (Becker et al., 2011; Kind et al., 2014). In addition, the flux distribution associated with this maximal yield, i.e. the 'optimal' pathway use, can be analyzed in more detail to investigate which pathways promote high yield metabolism (Melzer et al., 2009).

Furthermore, direct prediction of gene deletion, amplification and attenuation targets from elementary flux mode analysis is particularly useful in metabolic engineering for *in silico* strain design. For instance, identifying reactions, through a target potential coefficient (α), that show statistically significant correlation with a chosen objective, e.g. growth or product formation, delivers such targets (Melzer et al., 2009):

$$\alpha_{i,obj} = \frac{cov\ (v_{obj},v_i)}{\delta^2_{v_{obj}}} \qquad v_{obj} > 0 \quad \text{and} \quad 1 \leq i \leq q \qquad (\text{Eq. 4})$$

In this way, metabolic engineering targets for several industrial strains have been predicted and put into practice, delivering high-performance production processes for succinic acid by *Basfia succiniproducens* (Becker et al., 2013), lysine by *Corynebacterium glutamicum* (Becker et al., 2011) and poly-hydroxyalkanoates by *Pseudomonas putida* (Poblete-Castro et al., 2013). However promising, these approaches have so far not been applied to plant systems.

2.4.3. Isotopic labeling studies for flux analysis

In vivo metabolic fluxes can be determined by combining labeling studies with *in silico* approaches and fitting measured label distribution to a stoichiometric network model. In a typical labeling experiment for flux analysis in plants, a plant organ or tissue culture is exposed to a metabolic tracer, for a specific period of time (Fig. 2-7). This metabolic tracer contains one or more stable ^{13}C isotopes that are subsequently tracked throughout metabolism by nuclear magnetic resonance (NMR) or mass spectrometry (MS) (Ratcliffe & Shachar-Hill, 2006). From the experimentally determined isotope distribution pattern, fluxes can be calculated (Wittmann, 2002). Labeling studies can either be performed at isotopic steady-state, i.e. there are no observable changes in isotopic labeling pattern, by ^{13}C metabolic flux analysis (^{13}C-MFA) (Kruger et al., 2012) or at isotopic non-steady-state by isotopically non-stationary metabolic flux analysis (INST-MFA) (Young et al., 2008) or kinetic flux profiling (KFP) (Yuan et al., 2008). Although, ^{13}C-MFA is a well-established method that is particularly useful in heterotrophic tissue and cell suspension cultures, it is incompatible with photoautotrophic metabolism. Therefore, there is an increasing interest in deducing metabolic fluxes from time-dependent isotopic labeling patterns (Fernie & Morgan, 2013).

2.4.3.1. Isotopically instationary approaches for flux analysis of photoautotrophic metabolism

Two methods exist to process the obtained labeling time-courses from an isotopic non-steady-state tracer experiment. KFP relies on kinetic network models with detailed information about enzyme kinetics to describe the decaying metabolite enrichment (Yuan et al., 2008), whereas INST-MFA is a stoichiometry-based method that is capable of characterizing multiple isotopomers (Young et al., 2008). In both approaches, an iterative fitting to the measured isotopic labeling pattern, delivers an estimate of intracellular fluxes. Due to the underlying kinetics, KFP requires the additional determination of kinetic constants and pool sizes, which are both unnecessary for INST-MFA.

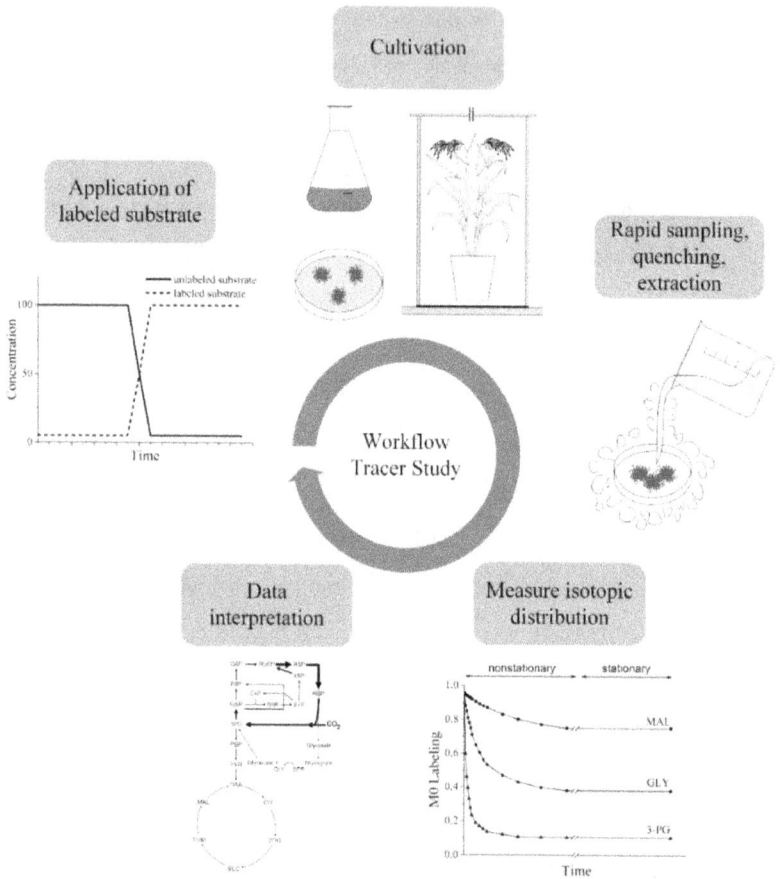

Figure 2-7: Workflow of label-based flux analysis

After a step-change from unlabeled to labeled substrate, the plant, plant part or cell culture is exposed to the tracer-substrate for a defined period of time. Rapid quenching and extraction ensure no metabolic conversions take place after sampling. Subsequently, the (transient) isotopic labeling pattern is determined by NMR or MS-based analytics, after which data interpretation can take place. Figure courtesy of L. Dersch, Institute of Systems Biotechnology, Saarland University.

The transient isotopic labeling patterns of metabolites from the central carbon pathways are typically used to infer fluxes in non-stationary approaches, as their high turnover rates are highly suitable. Therefore, it is necessary to take many samples immediately after the introduction of isotopic label, whereas sampling intervals can be increased towards later sampling time points. Additional labeling data from anabolic building blocks, such as amino acids, fatty acids, sugars or starch, could further be integrated into the analysis to improve flux identifiability (Ratcliffe & Shachar-Hill, 2006). Specially designed enclosures for the incubation with isotopically labeled CO_2 allow the study of specific plant parts such as leaves (Hasunuma et al., 2010; Schaefer et al., 1980) or whole plants (Tanaka & Osaki, 1983) and might include precise control of the growth environment (Andersen et al., 1961; Nouchi et al., 1995).

2.4.3.2. Algorithm behind ^{13}C-based metabolic flux analysis

Solving a minimization problem by fitting measured data to simulated values is at the heart of INST-MFA (Fig. 2-8 and Eq. 5). The objective is to find a set of intracellular metabolic fluxes (v) that best explain the measured data, thus that the difference between measured and simulated data is minimized in the following non-linear least squares problem:

$$\min \Phi = \min \mathrm{SSR} = \sum \frac{(x-x_m)^2}{\sigma_x^2} + \sum \frac{(c-c_m)^2}{\sigma_c^2} + \sum \frac{(r-r_m)^2}{\sigma_r^2} \qquad (\text{Eq. 5})$$

with $R \times v = r$ and $S \times v = 0$

Here, the variance-weighted sum of squared residuals (SSR) is a function of measured and predicted ^{13}C-labeling states (x), pool sizes (c) and rate measurements (r), subject to metabolite balances (Antoniewicz, 2015). This non-trivial and resource-intensive minimization problem, requires iterative algorithms due to its non-linear nature (Antoniewicz, 2015). Typically, an initial flux and pool size vector that complies to the mass balance equations is randomLy chosen, to simulate metabolite labeling with the isotopomer balance equations. Subsequently, the difference between simulated data and a set of experimentally determined measurements influences the direction and magnitude with which the initial flux vector is changed. With this optimized flux vector newly simulated labeling data are calculated and the process is repeated in an iterative way until the difference between simulated and observed data drops below a set threshold. If the then obtained flux distribution passes statistical fitness, it will be accepted as being an accurate representation of *in vivo* fluxes. An algorithm capable of iteratively adjusting fluxes and pool sizes and recalculating Φ until no further improvement is achievable, as described above, is the Levenberg-Marquardt local search algorithm (L-Ma) (Dennis & Schnabel, 1983; Madsen et al., 2004). Several software packages are readily available for steady-state flux estimation and subsequent statistical analysis, including 13CFLUX2 (Weitzel et al., 2013), FiatFlux (Zamboni et al., 2005), Metran (Yoo et al., 2008), OpenFLUX (Quek et al., 2009), FIA (Srour et al., 2011), influx_S (Sokol et al., 2012) and INCA (Young, 2014). INCA is currently the only publically available software additionally capable of INST-MFA.

2.4.3.1. Recent applications of photoautotrophic flux analysis

Prior to analysis in plants, fluxes of photoautotrophic metabolism were examined for the unicellular cyanobacterium *Synechocystis* by INST-MFA (Young et al., 2011). To date, only two studies in plants exist, which both estimated fluxes in whole *Arabidopsis* plants after transient labeling-experiments with $^{13}CO_2$. Szecowka et al. (2013) applied an extended KFP approach to wild type *A. thaliana* rosettes, whereas Ma et al. (2014) determined the flux phenotype of wild type and high-light acclimated leaves through INST-MFA. Both approaches were able to successfully resolve fluxes in a small carbon assimilation network. In conclusion, flux estimation tools are now available that could help in improving the knowledge-base for rational metabolic engineering of plant systems, as the interest in rational model-based engineering towards plants with improved performance is high (Cusido et al., 2014; Lotz et al., 2014; Shachar-Hill, 2013).

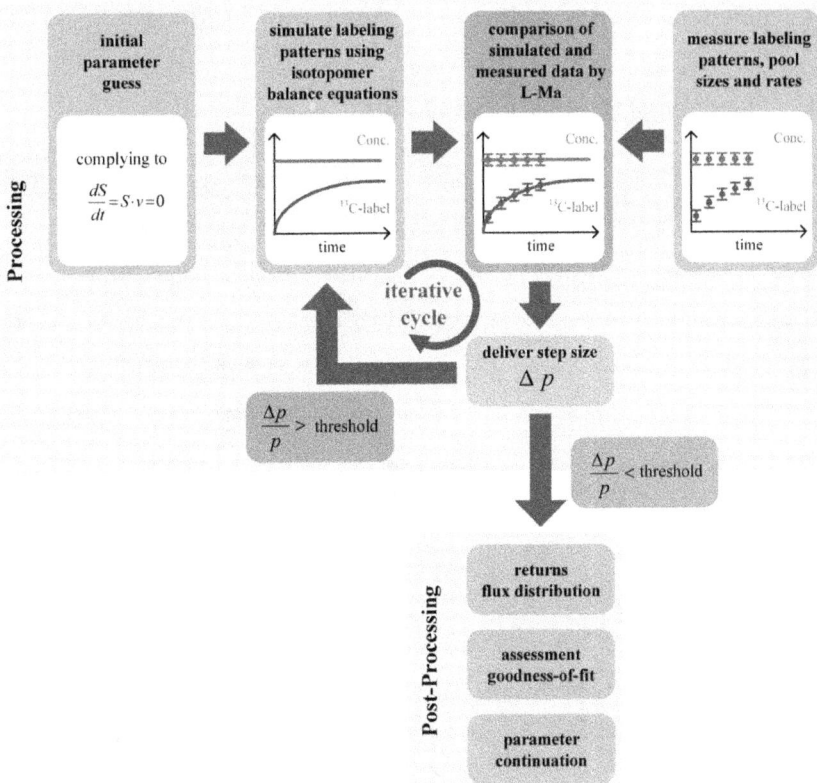

Figure 2-8: Overview on the basic functioning of the search algorithm to estimate fluxes from isotopically instationary ^{13}C labeling experiments

An initial parameter vector is used to simulate labeling patterns of internal metabolites using isotopomer balancing equations. The difference between simulated and measured data determines the subsequent step size and direction to optimize the starting vector by a Levenberg-Marquardt algorithm (L-Ma). In an iterative process the parameter set is improved until the parameters become static and the final flux distribution is delivered. If the flux set passes statistical post-processing, confidence intervals are determined with parameter continuation.

3.   Material and Methods

3.1.   *Arabidopsis thaliana* core metabolism

The initial draft of the central carbon metabolism of *A. thaliana* used for metabolic network reconstruction, considered curated knowledge collected from metabolic pathway databases: Kyoto Encyclopedia of Genes and Genomes (Kanehisa et al., 2008; KEGG, release 65.0), MetaCrop (Grafahrend-Belau et al., 2008; MetaCrop, release 2.0) and AraCyc (AraCyc, release 8.0; Mueller et al., 2003). This provided gross information on the genomic pathway repertoire. Where needed, the network was updated with experimental data and primary literature, as described in detail in chapter 4.1. These individual additions and specifications considered enzyme localization, cofactor usage, and intercompartmental transport.

3.2.   Computation of elementary flux modes

Elementary flux modes (EFMs) were calculated with efmtool, based on the null space approach and recursive enumeration with bit pattern trees (Terzer & Stelling, 2008). The EFM matrix, computed by the algorithm, comprises information on all thermodynamically and stoichiometrically possible pathways in the cell, which reduce metabolism into all feasible, unique, non-decomposable biochemical pathways (Schuster et al., 1999). Normalization of the EFM matrix and subsequent data interpretation was performed as described previously (Melzer et al., 2009). In short, relative fluxes were normalized to their respective substrate uptake and the theoretical biomass production of each elementary flux mode was determined. The contribution of different pathways to anabolic precursor formation was calculated by dividing the specific pathway flux into the precursor pool by the total flux into this metabolite. Subsequent flux correlation analysis investigated the target potential of individual metabolic reactions, i.e. biochemical conversions, not including transport reactions. If the statistical significance of the correlation was met, and the regression coefficient exceeded the cut-off of 0.7, the slope of the linear regression delivered the target potential coefficient for each individual reaction.

3.3. Integration of omics-data

3.3.1. Transcriptome

Experimental transcriptome data for integration with pathway fluxes were taken from a comprehensive study on the shift of gene expression between day and night in wild type *Arabidopsis* rosettes (Bläsing et al., 2005). In this work, the amplitude limit of gene expression (log2 value) during the diurnal cycle was measured with ATH1 arrays, and changes in gene expression were detected with a cut off value of 0.8. Here, genes encoding proteins of the central carbon metabolism were extracted from the data set and the underlying raw data were subsequently inspected to manually identify all genes, which exhibited a diurnal expression change, i.e. revealed on/off behavior or an unambiguous increase or decrease in expression during illumination and the opposite change during the dark period.

3.3.2. Fluxome

The metabolic fluxes of an illuminated *Arabidopsis* rosette (Szecowka et al., 2013) and of a heterotrophic *Arabidopsis* cell culture (Masakapalli et al., 2014), determined by KFP (kinetic flux profiling) and ^{13}C-MFA, respectively, were converted into (C-mol) (C-mol)$^{-1}$ to enable a straightforward comparison with the respective flux modes, obtained in this work, also given in (C-mol) (C-mol)$^{-1}$. A heterotrophic cell culture was chosen because, to our knowledge, heterotrophic whole plant flux studies have not been performed so far.

3.3.3. Quantum yield

The quantum yield of photosynthesis is derived from measurements of light intensity and rate of photosynthesis. The range of the quantum yield under ambient atmospheric conditions of *Arabidopsis*, as well as other C_3 plants, was taken from previous studies (Ehleringer & Bjorkman, 1977; Ehleringer & Pearcy, 1983; Long et al., 1993; Osborne & Garrett, 1983;

Skillman, 2008). Because not all harvested quanta are converted into chemical energy as some are lost through absorption by pigments, which are unable to contribute their excitation energy to photosynthesis, the experimental quantum yield values were corrected, assuming 47 % of photons are outside the photosynthetically active range (Hall & Rao, 1999). This provided a direct correlation between assimilated carbon dioxide and properly absorbed photons.

3.4.   Plant growth characteristics

The cultivation of rice seedlings was performed by Olaf Woiwode, Metanomics GmbH, Berlin.

Germination of *O. sativa* L. ssp. *japonica* Nipponbare seeds (CropDesign N.V., Zwijnaarde, Belgium) occurred in the dark on moist filter paper in a Petri dish (four days, 26 °C), after which the seeds were transferred into light for one day (500 µmol m^{-2} s^{-1} photosynthetically active radiation (PAR)). Prior to sowing, 0.7 dm^3 pots were saturated with deionized water and 0.15 % of the fungicide Proplant (Stähler, Stade, Germany). Hot water treatment (60 °C, 10 min) was applied to the rice seeds to prevent sheath rot. Sprouts were grown on soil (Einheitserde Type-GS90, 70 % organic fiber peat, 30 % clay, pH 5.5-6, Einheitserde- und Humuswerke Gebr. Patzer, Altengronau, Germany) under 13/11 h day/night cycles at an average irradiance of 500 µmol m^{-2} s^{-1} PAR during the light phase (Powerstar HQI-BT 400W, Osram, Munich, Germany). During cultivation, day/night temperatures of 26/21 °C and a relative humidity of 60 % were maintained. Top-irrigation took place two times a day with deionized water. Fifteen day old seedlings were used for the $^{13}CO_2$ labeling experiments.

3.5.   Imazapyr treatment

Herbicide treatment of rice seedlings was performed by Olaf Woiwode, Metanomics GmbH, Berlin and Lisa Dersch, Institute of Systems Biotechnology, Saarland University.

The herbicide Imazapyr, which belongs to the imidazolinone group, was applied to rice seedlings four hours prior to the $^{13}CO_2$ labeling experiments. A total concentration of 62.5 g ha^{-1} was reached by spraying the Imazapyr solution (0.3 mM Imazapyr (BASF SE, Ludwigshafen, Germany), 0.1 % (v/v) dimethyl sulfoxide (DMSO), 0.1 % (v/v) Dash® E.C. (BASF SE, Ludwigshafen, Germany)) on the plants with an airbrush. A solution containing 0.1 % (v/v) DMSO and 0.1 % (v/v) Dash® E.C was administered to the control plants.

3.6.   Isotopic labeling reactor

The design and construction of the flux incubator presented here, were carried out in collaboration with Detlev Rasch, Institute of Biochemical Engineering, Technical University Braunschweig.

The flux incubator used for isotopic labeling experiments with $^{13}CO_2$ consisted of a labeling reactor of approximately 620 L (75x75x110 cm), which was operated by a regulating unit that allowed air-humidification, regulation of temperature, ventilation, CO_2 regulation and turning of the integrated turning table (Fig. 3-1). The described flux incubator has previously been patented under WO 2014/079696.

Flux incubator unit. The incubator housing frame was constructed from duralumin (item Industrietechnik GmbH, Solingen, Germany). The use of polycarbonate (8 mm thickness, Hans Keim Kunststoffe GmbH, Rottweil, Germany) for the housing walls ensured spectral transmission of light. Solvent-free glue Loctite 406 (Henkel, Düsseldorf, Germany) and EPDM rubber seals (ethylene propylene diene monomer, Mercateo, Munich, Germany) guaranteed the incubator unit was gas-tight. EPDM rubber was used as sealing material, because of its high flexibility and endurance.

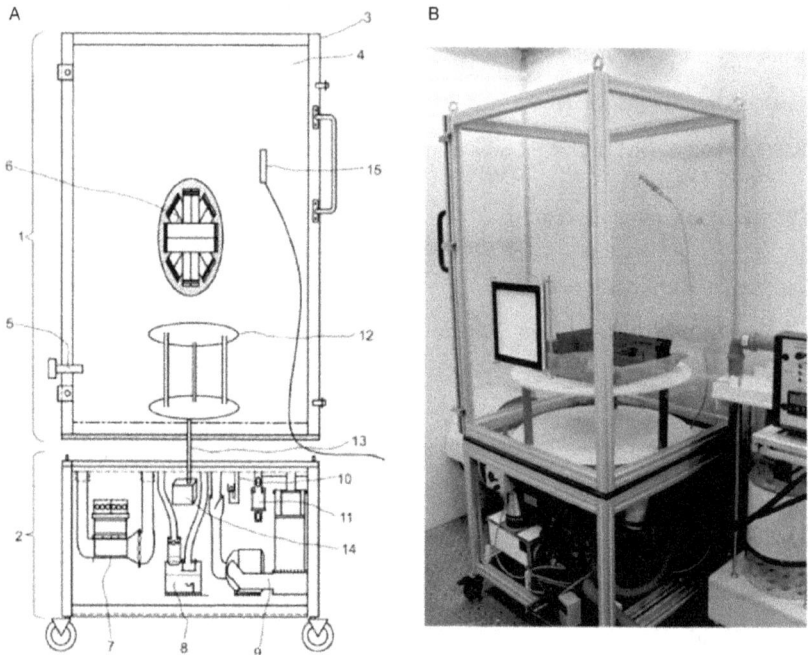

Figure 3-1: Tailor-made flux incubator for $^{13}CO_2$-labeling experiments of whole plants

Sketch (A) and photo (B) of the flux incubator, including all its components. 1. Flux incubator unit 2. Regulator unit 3. Housing frame 4. Housing walls 5. Injection port 6. Sampling port 7. Temperature regulating unit 8. Air humidification unit 9. CO_2 absorption unit 10. Pressure relief valve 11. Vacuum valve 12. Turning table 13. Drive axle 14. Motor for turning board 15. CO_2 measurement unit (Fuchs et al., 2014).

Temperature regulating system with integrated ventilation. The temperature in the flux

incubator was regulated by a peltier cooling/heating system (UETR-PT24V16A, uwe electronic

GmbH, Unterhaching, Germany). Two humidity-protected ventilators (170 m^3h^{-1}, 4312 NGN,

ebm-papst, Mulfingen, Germany) enabled circulation of air from the incubator through a peltier

cooling/heating element in the regulating unit (Fig. 3-2). The peltier element could be changed

from heating to cooling by changing the polarity, which was performed by the integrated

temperature regulator. After passing through the cooling/heating element, the air was humidified

and returned to the reactor.

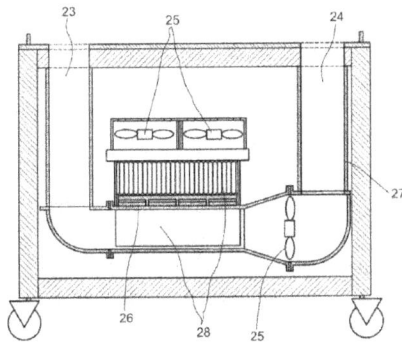

Figure 3-2: Temperature regulating system with integrated ventilation

Sketch of the temperature regulating system with integrated ventilation. 23. Air outlet 24. Air inlet 25.
Ventilator 26. Peltier element for heating/cooling 27. Flexible aluminum tube 28. Aluminum ribbed cooler
(Fuchs et al., 2014).

Air humidification system. The air humidification system (Fig. 3-3) comprised a water reservoir

with approximately 1.5 L deionized water and a submerged piezoceramic transducer (Conrad

Electronic SE, Hirschau, Germany). By mechanical oscillation of the ceramics (3 MHz), the

water in the container was nebulized. A second, smaller ventilator (5 m^3h^{-1}, 4312 NGN, ebm-

papst Mulfingen GmbH & Co. KG, Mulfingen, Germany) ensured air flow from the air outlet of

the temperature regulator through the humidifier back into the reactor (Fig. 3-3). In this way, the

airstream transported humidity into the flux reactor. A humidity regulator controlled the humidifier

(Conrad Electronic SE, Hirschau, Germany).

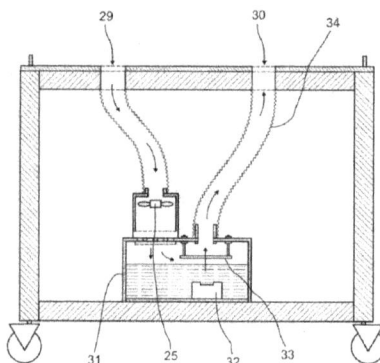

Figure 3-3: Air humidification system

Sketch of the air humidification system. 25. Ventilator 29. Air inlet 30. Air outlet 31. Container with water 32. Piezoceramics 33. Splash plate 34. Flexible tube (Fuchs et al., 2014).

CO_2 absorber. The CO_2 absorber was in place to remove atmospheric CO_2 from the flux reactor by pumping the air through soda lime pellets (Drägersorb® 800+, Mercateo, München, Germany) (Fig. 3-4 A). An electrical inflator (BRAVO 2000, Scoprega, Cassano D'Adda, Italy) drew 1800 L min^{-1} air from the flux reactor and pumped it through approximately 15 L of absorbing pellets. Soda lime pellets were chosen as absorber material because of their high CO_2 absorption capacity and incorporated quality indicator (color change). A back-pressure valve was in place to avoid back flow of CO_2 free air. The CO_2 depleted air was led over a fine particle filter to retain possible dust, before being recycled back into the flux reactor.

To avoid underpressure in the reactor during absorption, a vacuum valve with fine particle filter and CO_2 absorber material was integrated (Fig. 3-4 B). Also, to counteract the slight overpressure created right after switching off the inflator, a pressure relief valve was included. The combination of a vacuum and pressure relief valve allowed the flux incubator to be operated under normal pressure conditions.

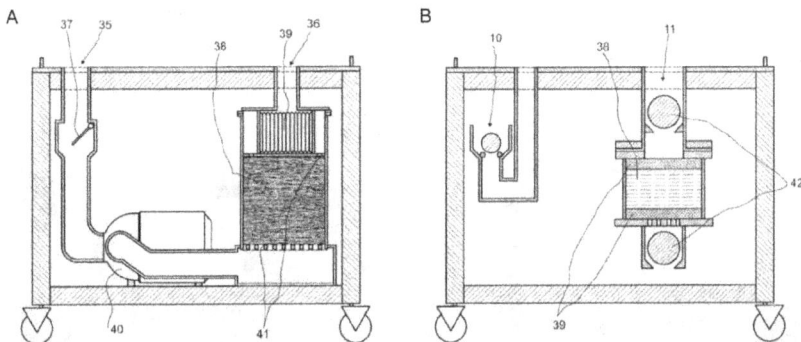

Figure 3-4: CO_2 absorbing system

Sketch of CO_2 absorber system consisting of a gas absorption unit (A) and a pressure compensation system (B). 10. Pressure relief valve 11. Vacuum valve 35. Air inlet 36. Air outlet 37. Back-pressure valve 38. Absorber material 39. Fine particle filter 40. Pump 41. Sieve 42. Ball valve (Fuchs et al., 2014).

Online mass spectrometry. To monitor and distinguish between the $^{12}CO_2$ and $^{13}CO_2$ levels in the flux reactor, an online mass spectrometry system (Pfeiffer Vacuum GmbH, Asslar, Germany) was installed. The HiQuad quadrupole mass spectrometer (100-240 V, 50/60 Hz) was able to detect molecules in a range between 1 and 512 amu.

Turning table. The polyvinylchloride turning table was operated by an external chopper transistor (Transistor-Gleichstromsteller Typ GS 24 S, EPH elektronik, Besigheim-Ottmarsheim, Germany) and could turn in both directions.

Sampling port. The sampling port (Fig. 3-5) consisted of an opening in one of the housing walls, covered with flexible rubber straps (THERABAND, The Hygenic Corporation, Akron, OH, USA). For sampling, a hand and specially designed sampling scissors could easily be moved between the rubber straps, without extensive gas-exchange taking place. During absorption the port was closed off with a 5 mm thick transparent polycarbonate plate (Hans Keim Kunststoffe GmbH, Rottweil, Germany). The polycarbonate plate was fixed to the sampling port by four strong magnets (Conrad Electronic SE, Hirschau, Germany), which were agglutinated to the plate with solvent-free glue Loctite 406 (Henkel, Düsseldorf, Germany).

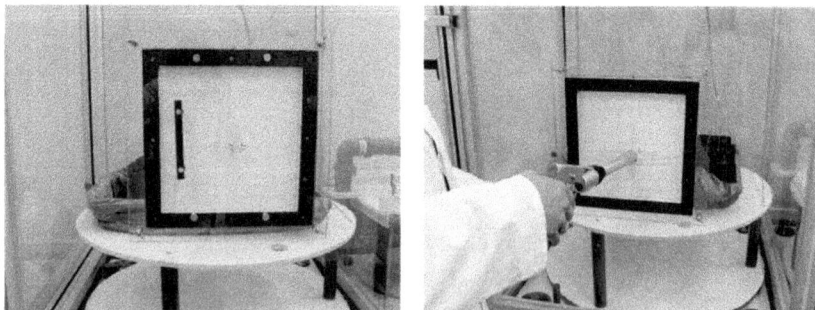

Figure 3-5: Sampling port

Picture of the sampling port that allowed sampling of individual plants throughout the $^{13}CO_2$ labeling experiments without perturbing the reactor atmosphere.

Sampling scissors. Gardening scissors (Classic Anvil Secateurs, GARDENA, Ulm, Germany), were modified for our purposes, by adding fitted foam rubber pieces (Meteor Gummiwerke, Bockenem, Germany) with solvent-free glue Loctite 406 (Henkel, Düsseldorf, Germany) to the blades (Fig. 3-6). During sampling, the foamed rubber fixed the cut seedling between the rubber pieces.

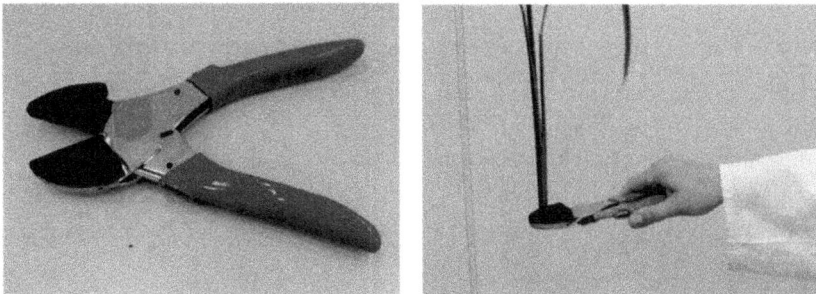

Figure 3-6: Sampling scissors
Picture of the tailor-made sampling scissors used during the $^{13}CO_2$ labeling experiments.

3.7.   Labeling of *Oryza sativa* seedlings with $^{13}CO_2$

The labeling of whole rice plants was performed in collaboration with Lisa Dersch, Institute of Systems Biotechnology, Saarland University, and Dr. Regine Fuchs and Marcel Schink from Metanomics GmbH, Berlin.

After placing 15-day old *O. sativa* seedlings (15 individual plants) in the flux incubator, atmospheric CO_2 was removed for 60 seconds by soda lime pellets in the CO_2 absorber unit. Following absorption, $^{13}CO_2$ (> 99 atom% ^{13}C, Eurisotop, Saarbrücken, Germany) was injected with a gas-tight syringe (500 mL, Hamilton Company, Reno, NV, USA) through a septum in the housing wall of the flux incubator, so that the reactor atmosphere contained 400 ppm $^{13}CO_2$. Throughout the 30 minute lasting experiment, all 15 plants were harvested. The first sample was collected immediately after injection (0 seconds), whereas the subsequent time points were taken after 10, 20, 30, 40, 50, 60, 90, 120, 150, 180, 300, 420, 600 and 1800 seconds. The specially designed sampling scissors enabled ultra-quick (1-2 seconds) harvesting of the entire above-ground shoot through the sampling port, followed by instantaneous quenching in a liquid nitrogen bath. Furthermore, immediately following the above-ground harvest at 300, 600 and 1800 seconds, roots were plucked from the soil. These roots were washed in deionized water and both the seed and remaining green tissue were removed, before freezing in liquid nitrogen. In total 60 seedlings were harvested for each condition, constituting 5 replicate experiments.

3.8. Analytics of metabolite labeling and concentration

Analytical processing of samples was performed at Metanomics GmbH, Berlin, in collaboration with Lisa Dersch, Institute of Systems Biotechnology, Saarland University.

3.8.1. LC-MS/MS

Extraction was performed as described earlier (Balcke et al., 2011) with some alterations: The samples (5 mg) were extracted with 150 µL 1.5 M ammonium acetate. Extraction and centrifugation (0.22 µm Millipore filter, Merck Millipore, Billerice, MA, USA) were performed with 300 µL of the polar phase. The centrifugal filter units were washed with 200 µL HPLC grade water. The following sample preparation steps were not changed.

Mass isotopomer distribution was determined by LC-MS/MS for G6P, PEP, PYR, 6PG, FBP, S7P, 2PG, 3PG, F6P, P5P, GLYCO and RBP (For abbreviations see Appendix 10.5.2). All possible isotopomers were added to the method of Balcke et al. (2011).

Pool sizes were determined quantitatively for PEP, P5P, S7P, 2PG, 3PG, E4P and RBP (For abbreviations see Appendix 10.5.2). For each individual metabolite an external calibration series was used in the corresponding calibration range (1 – 100 000 ng/mL). To correct for ion suppression, samples and calibration levels were spiked with 50 µL U^{13}C-yeast extract ISTD.

3.8.2. GC-MS

Lyophilized tissue (5 mg) was used for metabolite profiling. Metabolites were extracted with the use of accelerated solvent extraction with polar (methanol 80 % v/v in water) and non-polar (methanol 40 % v/v in dichloromethane) solvents. Subsequent analyses of metabolites by gas chromatography–mass spectrometry (GC-MS) were performed as described elsewhere (Roessner et al., 2000; Walk, 2007). AKG, SUCC, FUM, alanine, valine, leucine, isoleucine, glycine, proline, serine, threonine, aspartate/asparagine, methionine, glutmate/glutamine, tyrosine, lysine, phenylalanine, inositol, glucose and fructose were measured quantitatively, whereas the mass isotopomer distributions of AKG, SUCC, FUM, CIT, MAL, PYR, GLYCER, G6P, sucrose, alanine, aspartate, glutamate, isoleucine, leucine, phenylalanine, proline, valine, serine and threonine were determined qualitatively (For abbreviations see Appendix 10.5.2).

3.8.3. GC-irMS

Sample preparation for GC-C-irMS analysis was performed according to the procedures of Hurkman and Heinzle (Heinzle et al., 2008; Hurkman & Tanaka, 1986). Deep-frozen, ground plant material (5 mg) was extracted with an extraction buffer (0.175 M tris(hydroxymethyl)aminomethane/hydrochloric acid, pH 8.8, 5 % sodium-dodecylsulfate, 15 % glycerol, 0.3 M dithiothreitol). Proteins were precipitated with acetone and hydrolyzed with 6 M hydrochloric acid at 100 °C. Amino acids were derivatized with Pyridin and MTBSTFA (N-methyl-N-tert-butyldimethylsilyl-trifluoroacetamide, Macherey-Nagel, Düren, Germany) at 60 °C for 1 h.

Enrichment with ^{13}C labeling was measured on a GC–C–irMS instrument (Thermo Fisher Scientific, Bremen, Germany) comprising a Trace 1300 gas chromatograph, a TriPlus RSH autosampler, a GC Isolink (with a NiO combustion tube in combination with NiO and CuO wires set at 1000 °C), a ConFLO IV interface and a Delta V Advantage isotope ratio mass spectrometer. H_2O generated in the combustion reactor was removed by passage of the combustion products through a water permeable Nafion membrane. The Agilent 6890 gass chromatograph (splitless, injection volume 1 µL, flow rate 2.3 mL/min, oven temperature: 1 min at 70°C, 8°C/min to 280°C, 50°C/min to 340°C, 3 min at 340°C, Agilent Technologies, Santa

Clara, CA, USA) with a DB-5 column (30 m, 0.25 mm I.D., 0.25 µm film thickness, Agilent

Technologies, Santa Clara, CA, USA) with helium as carrier gas.

Sample preparation for EA-C-irMS analysis and C/N elemental analysis. Finely ground dry

plant material (5 mg) was weighed into tin capsules (3.3 x 5 mm, HEKAtech GmbH, Löbau,

Germany) prior to analysis. Both ^{13}C labeling measurements and C/N-ratio measurements were

performed on a FLASH HT Plus elemental analyzer (Thermo Fisher Scientific, Bremen,

Germany) run in the single reactor combustion mode and operated at 1020 °C. The dedicated

reactor sections comprised layers of Cr_2O_3 for combustion, reduced Cu for reduction and

silvered cobaltous / cobaltic oxide to remove halogens and sulfur.

Isotope ratio measurement, calculation of $\delta^{13}C$ and calculation of C/N elemental

composition. Three Faraday cup collectors for m/z 44, 45, and 46 were used for detection of

CO_2. Isotope ratios were calculated from the relative abundances of these mass traces by

integrating their ion currents. Isotope ratios were calibrated against reference CO_2 of known

isotopic composition, which was introduced directly into the ion source five times at the

beginning of every run. The reference gas was calibrated against Vienna Pee Dee Belemnite

(VPDB) scale (Coplen et al., 2006). δ-values were corrected for natural labeling of the

derivatization agent by the method described by Heinzle et al. (2008). Natural labeling of both

oxygen atoms was corrected for by applying the method provided in INCA (Young, 2014). C/N

elemental composition calculations were performed with Isodat 3.0 software (Thermo Fisher

Scientific, Bremen, Germany) with acetanilide being used for calibrating the instrument. The

software allows calculating C/N elemental composition both based on the data of the integrated

thermal conductivity detector (TCD) and on the mass spectral data. In the latter case, N_2 is

measured with the Faraday cups set at m/z 28, 29, and 30. Enrichments were measured for alanine, aspartate, asparagine, glutamate, glutamine, glycine, lysine, serine, tyrosine, valine, sucrose, glucose, fructose, inositol and malate.

3.8.4. Amino acid composition

Amino acid analysis was performed with adaptions based on the proprietary AccqTag analysis kit (Waters Corp, Milford, MA). Lyophilized plant material (4.5 mg) was hydrolyzed with 450 µL 6N HCl including 45.8 mg/L norvaline as ISTD and 50 µL 100 mM sodium dithionite for 24h at 110°C. A 50 µL aliquot was transferred and dried under vacuum, derivatized with 80 µL borate buffer and 20 µL kit reagent solution as supplied by the kit for 15 min at 55°C. The sample was filtrated and subjected to UPLC-UV analysis as described by the kit's instructions.

3.8.5. Starch and sucrose analysis

Sucrose and starch levels were determined in 10-20 mg homogenized and lyophilized leaf material by ball milling (MM300 from RETSCH, Haan, Germany) for 5 min at 30 Hz in 1.5 mL 80 % (v/v) ethanol/water. After cooling to 4°C, the samples were separated by centrifugation (5 min, 9000 xg) into pellet and supernatant. Of the supernatant, 200 µL were dried down. The dried residue containing free sugars was redissolved and derivatized for 90 min at 60°C with methoxyamine hydrochloride in pyridine, followed by a 30 min treatment with MSTFA (Rosner AG, Safenwil, Swiss) at 60°C. The pellet was washed in 1.5 mL extraction solution and dried down. Subsequently, the dried pellet was incubated at 55°C with 850 µL buffer (100 mM imidazol and 5 mM $MgCl_2$ at pH 7), 75 µL amylase and 75 µL amyloglucosidase on a shaker for 24 h. In this way, starch was broken down in to its monomeric subunits. After centrifugation, 100 µL of the supernatant was analyzed for its sugar content as described above.

Quantitation of sugar concentration was performed using an Agilent MSD instrument. The Agilent 6890 gass chromatograph (split 15:1, injection volume 0.5 µL, flow rate 1.3 mL/min, oven temperature: 1 min at 70°C, 120°C/min to 120°C, 18°C/min to 220°C, 45°C/min to 350°C, 2 min at 350°C, Agilent Technologies, Santa Clara, CA, USA) with a RXI-XLB column (20 m, 0.18 mm

I.D., 0.18 µm film thickness, Agilent Technologies, Santa Clara, CA, USA) with helium as carrier gas. Quantitative measurements of the compounds of interest were performed using an external calibration.

3.8.6. Fatty acid methyl ester analysis

Lipids were extracted from 25 mg homogenized and lyophilized leaf material by ball milling (MM300 from RETSCH, Haan, Germany) for 5 min at 30 Hz in 1000 µL methyl *tert*-butyl ether (MTBE). After centrifugation (5 min, 9000 x*g*) 200 µL supernatant were mixed with 10 µL TMSH-solution (0.25 mM in Methanol) and incubated for 20 minutes at room temperature. Fatty acid methyl esters were quantified gass chromatographically on an Agilent 6890 gass chromatograph (split ratio: 1:75, 0.5 mL/min flow rate, oven temperature: 140°C, 0.5 min, 20°C/min to 240°C, 2.5 min, Agilent Technologies, Santa Clara, CA, USA) with a MXT®-WAX column (10 m x 0.18 mm x 0.2 µm , Restek Corporation, Bellefonte, PA, USA) and helium as carrier gas.

3.9.   *Oryza sativa* core metabolic network

A comprehensive metabolic isotopomer network for *O. sativa* was carefully reconstructed based on curated knowledge collected from metabolic pathway databases: Kyoto Encyclopedia of Genes and Genomes (Kanehisa et al., 2008; KEGG, release 65.0), MetaCrop (Grafahrend-Belau et al., 2008; MetaCrop, release 2.0) and Plant Metabolic Network (OryzaCyc, version 2.0). In this manner, a first layout of the genomic pathway repertoire was generated, which was subsequently enriched with information from primary literature. These individual additions and specifications considered enzyme localization, cofactor usage, and intercompartmental transport as described in detail in chapter 4.2. Finally, before deciding on the final network topology, different possible scenarios were tested *in silico* as described in Chapter 6.3.

3.10. *In vivo* flux estimation by INST-MFA

The estimation of intracellular metabolic fluxes by isotopically non-stationary ^{13}C metabolic flux analysis was performed in Isotopomer Network Compartmental Analysis (INCA) (Young, 2014), ran in MATLAB (R2012b, The Mathworks Inc., Natick, MA, USA). The Levenberg-Marquardt algorithm implemented in INCA, varied the metabolic fluxes and pool sizes, while minimizing the sum of squares error between experimentally determined and simulated mass isotopomer distributions (MIDs) (Young et al., 2008). The relative convergence tolerance, which determines the termination threshold based on the size of Δp and the current parameter vector p, was set to 5 %. Tau, controlling the initial size of the damping parameter, was set to 1e-05. Fifty individual parameter estimation runs from random initial values yielded the best-fit estimate.

Statistical evaluation assessed goodness-of-fit by a χ^2-test with n-p degrees of freedom (DOF), where n is the number of independent measurements and p is the number of fitted parameters. The expected SSR range was calculated as $\left[\chi^2_{\alpha/2}(n-p), \chi^2_{1-\alpha/2}(n-p)\right]$, where $\alpha = 0.05$ is a threshold p-value at which the fit is rejected. Additionally, normality of the error-weighted residuals was evaluated with a Lilliefors test at the previously described α significance level. Parameter continuation was performed to determine the 95 % confidence intervals on net and exchange fluxes, as well as on pool sizes (Antoniewicz et al., 2006).

Results and Discussion

4.   Metabolic Network Reconstruction

4.1.   Genome-based metabolic network of *Arabidopsis thaliana*

As first focus, a large-scale metabolic network of *Arabidopsis thaliana's* core carbon pathways was carefully reconstructed, to provide a solid basis for subsequent simulation studies by *in silico* elementary flux mode analysis. In total, the network described 483 metabolic conversions and represented all relevant physiological characteristics of a plant cell, such as anabolism, catabolism, energy and redox supply. This final network accounted for 1567 metabolic genes, known to exist in *A. thaliana* (AraCyc, release 8.0).

Core metabolism. The created network contained the glycolytic Embden-Meyerhof-Parnas (EMP) pathway, the pentose phosphate (PP) pathway, the tricarboxylic acid (TCA) cycle, the Calvin-Benson-Bassham (CBB) cycle for photosynthesis, photorespiration, starch biosynthesis and degradation, energy metabolism as well as anabolic pathways for biomass synthesis (Fig. 4-1). The latter considered compartment-specific supply of individual precursors (Appendix 10.1 and 10.2) (de Oliveira Dal'Molin et al., 2010a; Mintz-Oron et al., 2012). As natural carbon sources, internal starch and atmospheric CO_2 were included as substrates, respectively.

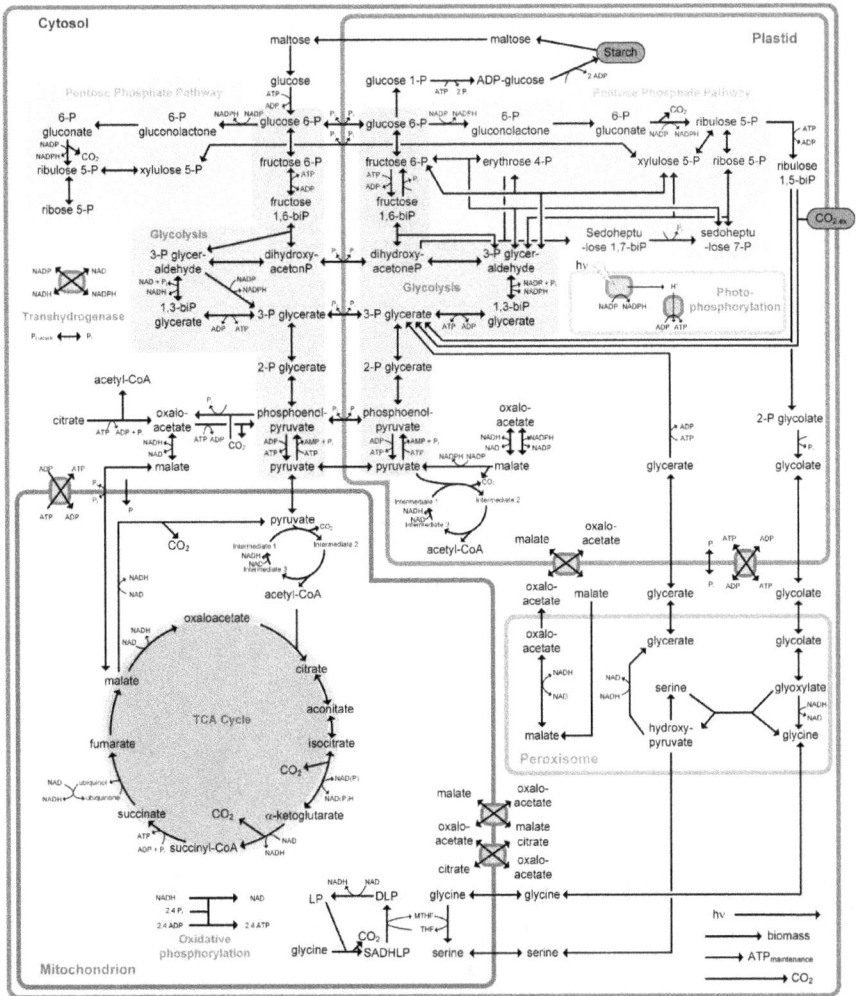

Figure 4-1: Metabolic network for an Arabidopsis leaf

Metabolic network for Arabidopsis leaf delivering a compartmented view on central carbon metabolism, including the distribution of carbon core metabolic pathways across four compartments: cytosol, plastid, mitochondrion and peroxisome. The purple transmembrane transporters mediate antiport of the selected metabolites. Uni- and bidirectional transport is represented by uni- and bidirectional arrows across the membrane, respectively. CO_2 is allowed to freely diffuse throughout the cell, whereas photons can pass both the cytosolic and plastidic membranes. A more detailed photophosphorylation can be found in Fig. 4-2. For visualization purposes, anabolism is simplified to a single biomass forming reaction.

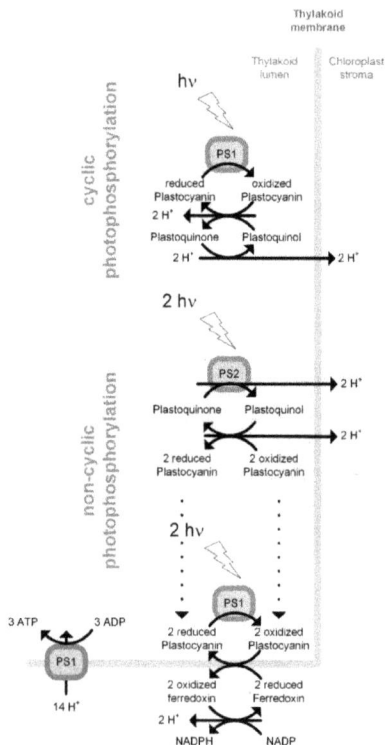

Figure 4-2: Cyclic and non-cyclic photophosphorylation

A plasticity in cyclic and non-cyclic photophosphorylation exists, however, a standard stoichiometric ratio of cyclic and non-cyclic photophosphorylation of 2:1 was assumed for modeling. This ratio accounts for 2x4 photons for non-cyclic electron flow and 1 absorbed photon for cyclic photophosphorylation.

Compartmentation. The metabolic network reflected the four central compartments in plants: cytosol, peroxisome, mitochondrion and plastid (AraCyc, release 8.0; Fernie et al., 2004; Kruger & von Schaewen, 2003; MetaCrop, release 2.0; Orzechowski, 2008; Plaxton, 1996; Schnarrenberger & Martin, 2002; Sweetlove & Fernie, 2013). The cytosol comprised the reactions of the EMP pathway (Plaxton, 1996), the oxidative part of the PP pathway (Kruger & von Schaewen, 2003) and reactions of starch degradation from maltose (Orzechowski, 2008), respectively. The plastid contained the photosynthetic CBB cycle, a second copy of the EMP pathway (Plaxton, 1996), the oxidative and the non-oxidative PP pathway (Kruger & von Schaewen, 2003) and starch metabolism (Orzechowski, 2008). The TCA cycle was assigned to the mitochondrion (Schnarrenberger & Martin, 2002). The photorespiratory system, known to be a multi-compartment process, was distributed accordingly across the plastid, the peroxisome and the mitochondrion (Raghavendra et al., 1998). Additionally, each compartment contained malate dehydrogenase (MetaCrop, release 2.0; Visser et al., 2007). Furthermore, pyruvate kinase was considered as cytosolic and as plastidic reaction (Plaxton, 1996), whereas the pyruvate dehydrogenase complex was assigned to mitochondrion and plastid, respectively (Schnarrenberger & Martin, 2002). The supply of cytosolic acetyl-CoA was attributed to ATP-citrate lyase in the cytoplasm, which uses citrate as a substrate (Fatland et al., 2005). Malic enzyme, specific for photosynthetic tissues, was assigned to the plastid (Plaxton, 1996).

Intercompartmental and external transport. The separation of metabolic routes in distinct organelles, requires translocation of specific compounds across the cellular membranes. Based on previous experimental evidence, uni- or bi-directional transport between cytosol and mitochondrion was assumed for pyruvate, malate, inorganic phosphate, glycine and serine, respectively, whereas antiporters were considered for malate/oxaloacetate, citrate/oxaloacetate and ATP/ADP (Grafahrend-Belau et al., 2008; Haferkamp et al., 2011; Klingenberg, 2008; Laloi, 1999; Picault et al., 2004). Additionally, transport between cytosol and plastid was considered for 3-phosphoglycerate, glycerate, glycolate, malate/oxaloacetate, pyruvate, phosphoenolpyruvate, xylulose 5-phosphate, glucose 6-phosphate, dihydroxyacetonephosphate, maltose, ATP/ADP and inorganic phosphate, respectively (Eicks et al., 2002; Facchinelli & Weber, 2011; Fischer, 2011; Furumoto et al., 2011; Grafahrend-Belau et al., 2008; Kruger & von Schaewen, 2003). Hereby, the translocation of phosphorylated carbohydrates across the plastidial membranes occurred by simultaneous antiport of inorganic phosphate (Fischer, 2011). In addition, active peroxisomal membrane transfer of malate/oxaloacetate, glycerate, glycolate, glycine and serine occurred (Raghavendra et al., 1998; Visser et al., 2007). So far, a transporter for acetyl-CoA has not been discovered and was therefore not incorporated (Schwender et al., 2006). CO_2 was assumed to freely diffuse within the cell (Kaldenhoff et al., 2014), photons were capable of penetrating both the extracellular and plastidic membranes (Terashima et al., 2009) and inorganic phosphate was available from the vacuole (Rausch & Bucher, 2002). Furthermore, the applied algorithm required *in silico* external exchange reactions for unbalanced metabolites, such as biomass, maintenance ATP, photons and starch. In addition, the exchange of reducing equivalents between the cytosol and both the mitochondrion and the peroxisome was mediated by malate dehydrogenase coupled to malate and oxaloacetate channeling across the organelles barrier.

Energy household. Redox, energy and phosphate metabolism were compartmentalized across the different organelles. This included confinement of the photosynthetic light reactions to the plastid and oxidative phosphorylation to the mitochondrion. In the vacuole, a storage pool for inorganic phosphate was considered (Rausch & Bucher, 2002). From here, inorganic phosphate could be transported throughout the cell by specific carriers between the cytosol and both the plastid and the mitochondrion. ATP/ADP antiporters were incorporated in the mitochondrial and plastidic membrane. In addition, the exchange of reducing equivalents between the cytosol and the plastid, the mitochondrion and the peroxisome was mediated by malate dehydrogenase coupled to malate and oxaloacetate channeling across the organelles barrier. To account for widely abundant isoenzymes, capable of utilizing either NADPH or NADH, or both molecules as cofactor, an oxidoreductase for interconversion of NADPH and NADH was included in the cytosol (Cheung et al., 2013). Although more realistic, additional simulation of a network without compartment-specific energy, redox and phosphate metabolism, showed that the efficiency of biomass formation was not affected by the introduced compartmentalized energy, redox and phosphate acquisition (data not shown). The following considerations were included for the energy efficiency. The ratio of ATP:NADPH produced in a photosynthetic cell depends on a generally accepted plasticity of the photosynthetic light reactions for energy production (Fig. 4-2) (Allen, 2003; Cruz et al., 2005; Foyer et al., 2012; Kramer et al., 2004; Munekage et al., 2008). As it is still unresolved how this mechanism functions exactly, an average ATP to NADPH ratio of 1.5, which accounted for cyclic electron flow around photosystem 1 of one photon and a non-cyclic electron flow caused by 8 photons (de Oliveira Dal'Molin et al., 2010a). In certain simulations, the photosynthetic plasticity was investigated by varying the ratios between cyclic and non-cyclic electron flow. Oxidative phosphorylation was incorporated in the mitochondrion. In order to provide sufficient ATP for maintenance, a surplus of ATP was set as constraint for the modeling (Krömer et al., 2006).

Anabolic pathways for biomass synthesis. The biochemical composition of *Arabidopsis* leaves was carefully collected from selected publications (Appendix 10.1). Nearly half of the carbon is stored in the cell wall (Herrero et al., 2013; Reiter et al., 1997; Zablackis et al., 1995), whereas one third was contained in proteins (Tschoep et al., 2009) (Fig. 4-3 A). Based on experimental data, the remaining carbon was distributed among lipids (Fan et al., 2013; Shen et al., 2010; Stahl et al., 2004), carbohydrates (Reiter et al., 1997; Saxena et al., 2013; Tschoep et al., 2009), porphyrines (Nowicka et al., 2009; Tardy & Havaux, 1996) and other biomass components (Arnqvist et al., 2008; Murray & Thompson, 1980; Suzuki et al., 2004; Tschoep et al., 2009). Together, the anabolic pathways synthesizing these biomass building blocks, form the most important carbon sink during growth. As most of these peripheric biosynthetic pathways are strongly linear and a metabolic steady-state is assumed, they can easily be summarized into a single, lumped biomass equation, starting from 12 central metabolic precursor metabolites

(Appendix 10.2). The localization of the individual enzymes determines where a certain biomass

component is synthesized and thus, from which organelle its precursor originates. For instance

aromatic amino acids are synthesized from phosphoenolpyruvate and erythrose 4-phosphate in

the plastid (Tzin & Galili, 2010), whereas cellulose is known to originate from hexose

6-phosphate in the cytosol (McFarlane et al., 2014). This in turn, determines which compartment

contributes most to providing carbon for growth (Fig. 4-3 B). As most carbon is needed for the

synthesis of cell wall components, which originate from pathways located in the cytosol, it is not

surprising that the cytosol provides more than 50 % of the carbon for growth.

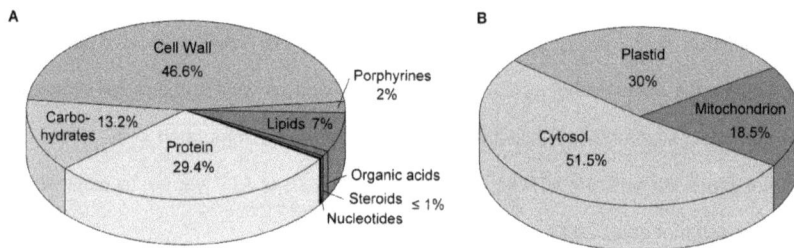

Figure 4-3: Biomass composition of A. thaliana leaves

(A) Carbon content-normalized biomass composition of A. thaliana leaves, based on the information given in Appendix 10.1 (expressed in %) (B) Compartment-specific carbon demand for growth of A. thaliana leaves (expressed in %), which was calculated by combining the stoichiometric precursor demand (Appendix 10.2) of each biomass constituent with its relative composition (Appendix 10.1).

By lumping the anabolic pathways towards biomass synthesis, the equation system to describe

Arabidopsis' metabolism could be represented in a compact form for more efficient calculation

by 129 reactions and 70 metabolites, while remaining descriptive of the entire set of metabolic

conversions (Appendix 10.3). Hereby, 34 reactions belonged to intercompartmental and

extracellular transport, respectively. Based on thermodynamic properties provided in the above-

mentioned databases, 72 reactions were constrained as irreversible. This now allowed the computation of all feasible metabolic routes in the leaf, through enumeration of elementary flux modes.

Pathways to valuable plant traits. Selected scenarios investigated the plant potential to synthesize iso-pentenyl pyrophosphate (IPPP), the building block for terpenes and terpenoids, which is provided via the mevalonate pathway or the methyl-erythritol pathway (Rodríguez-Concepción & Boronat, 2002). Additionally, different amino acids and carbohydrates were included in the analysis. L-Tryptophan, L-lysine, L-threonine, L-methionine and L-cysteine exhibit nutritional value (Amir, 2010). Tryptophan is synthesized through the shikimate pathway (Zhao & Last, 1995) and lysine synthesis uses the diaminopimelate pathway (Jander & Joshi, 2009). Threonine and methionine are obtained from L-aspartate and share the common intermediate L-homoserine (Jander & Joshi, 2009). Methionine and cysteine biosynthesis involves sulfate assimilation (Wirtz & Hell, 2006). Additionally, L-proline was included as important factor in salt and drought tolerance (Chen et al., 2007), which is synthesized from L-glutamate (Roosens et al., 1998). Furthermore, the carbohydrates sucrose and starch were investigated, because of their dietary value, and lignocellulose was considered due to its impairing effect in biorefineries (Dey & Harborn, 1997; Hasunuma et al., 2013; Humphreys & Chapple, 2002; Reiter, 2002; Streb & Zeeman, 2012).

4.2.   Large-scale isotopomer network of *Oryza sativa*

To obtain realistic predictions of *in vivo* metabolism, it is crucial that ^{13}C-INST-MFA calculations rely on a dependable and meticulously formulated metabolic network. For this purpose, a large-scale isotopomer model of central carbon metabolism was constructed for *Oryza sativa* seedlings.

Core metabolism. The isotopomer model of *O. sativa* contained information about both reaction stoichiometry and carbon transitions. It consisted of the glycolytic Embden-Meyerhof-Parnas (EMP) pathway, the pentose phosphate (PP) pathway, the tricarboxylic acid (TCA) cycle, the Calvin-Benson-Bassham (CBB) cycle for photosynthesis and photorespiration. Additionally, sucrose export to the root and individual pathways for amino acid and fatty acid metabolism were included. The remaining anabolic pathways were summarized in the biomass synthesis reaction (Fig. 4-4). The latter considered compartment-specific supply of individual precursors for cell wall, nucleotide and pigment synthesis, as well as provision of fatty acids, amino acids, organic acids, soluble sugars and starch (Appendix 10.1 and 10.2). Atmospheric CO_2 was the sole carbon source, which was allowed to be re-assimilated after respiration (Zabaleta et al., 2012). The final network consisted of 75 reactions and 65 metabolites, whereby 5 reactions belonged to intercompartmental and extracellular transport, respectively (Appendix 10.3). Based on thermodynamic properties provided in the enzyme database BRENDA (BRaunschweig ENzyme DAtabase), 56 reactions were constrained as irreversible.

Figure 4-4 (next page): Metabolic network for *Oryza sativa* seedling

The isotopomer metabolic model visualized here, includes the central carbon metabolism, anabolic precursor demand and export of sucrose to root metabolism.

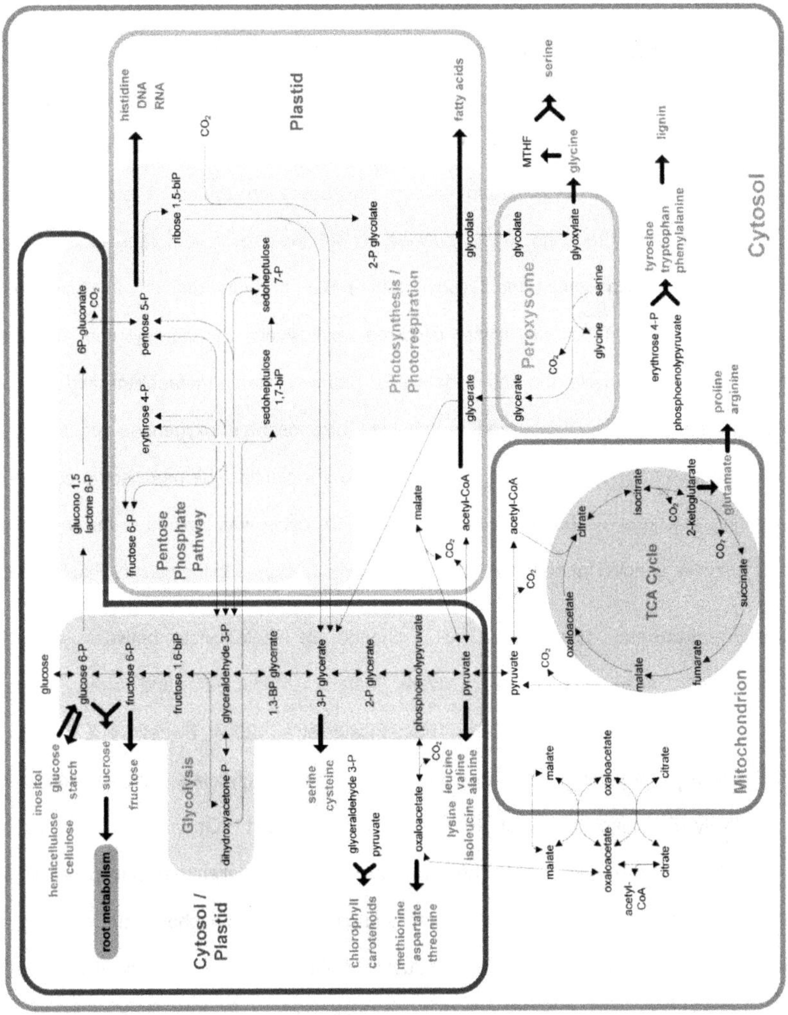

Cytosol
Plastid
Cytosol / Plastid
Mitochondrion
Peroxysome
Pentose Phosphate Pathway
Photosynthesis / Photorespiration
Glycolysis
TCA Cycle
root metabolism
glucose
glucose 6-P
fructose 6-P
fructose 1,6-biP
glyceraldehyde 3-P
dihydroxyacetone P
1,3-BP glycerate
3-P glycerate
2-P glycerate
phosphoenolpyruvate
pyruvate
glucono 1,5 lactone 6-P
6P-gluconate
pentose 5-P
erythrose 4-P
ribose 1,5-biP
sedoheptulose 1,7-biP
sedoheptulose 7-P
2-P glycolate
glycolate
glycerate
acetyl-CoA
malate
glyoxylate
serine
glycine
MTHF
histidine
DNA
RNA
fatty acids
inositol
hemicellulose
cellulose
glucose
starch
sucrose
fructose
serine
cysteine
glyceraldehyde 3-P
pyruvate
chlorophyll
carotenoids
oxaloacetate
methionine
aspartate
threonine
lysine
isoleucine
leucine
valine
alanine
citrate
isocitrate
2-ketoglutarate
glutamate
succinate
fumarate
proline
arginine
erythrose 4-P
phosphoenolpyruvate
tyrosine
tryptophan
phenylalanine
lignin
acetyl-CoA

Compartmentation. The compartmented nature of plant cells was accounted for by organizing the large-scale metabolic network into four compartments: cytosol, peroxisome, mitochondrion and plastid (Fernie et al., 2004; Kruger & von Schaewen, 2003; MetaCrop, release 2.0; OryzaCyc, version 2.0; Orzechowski, 2008; Plaxton, 1996; Schnarrenberger & Martin, 2002). The reactions of the EMP pathway and the oxidative part of the PP pathway occurred both in the plastid and the cytosol (Kruger & von Schaewen, 2003; Plaxton, 1996), which dramatically increased the number of possible carbon metabolization routes. Therefore, an artificial compartment was introduced to the network to enhance resolution, representing a merged cytosolic and plastid compartment (See chapter 6.3 for more details). Furthermore, sugar metabolism and ATP-citrate lyase occurred exclusively in the cytosol, whereas the photosynthetic CBB cycle, the non-oxidative PP pathway, starch metabolism, and a part of the photorespiratory system, catalyzed by ribulose bisphosphate oxygenase, phosphoglycolate phosphatase and glycerate kinase, were unique to the plastid. The peroxisome comprised the remaining reactions of photorespiration and the TCA cycle was located in the mitochondrion. Malic enzyme, specific for photosynthetic tissues, was assigned to the plastid (Plaxton, 1996).

Intercompartmental transport. The exchange of metabolites between cytosol and mitochondrion was assumed for pyruvate, malate, oxaloacetate and citrate, respectively (Grafahrend-Belau et al., 2008; Laloi, 1999; Picault et al., 2004; Sweetlove & Fernie, 2013). Plastid-cytosol transport was considered for glycerate, glycolate, malate and pyruvate, respectively (Facchinelli & Weber, 2011; Furumoto et al., 2011; Grafahrend-Belau et al., 2008; Kruger & von Schaewen, 2003). The transport of 3-phosphoglycerate, phosphoenolpyruvate, xylulose 5-phosphate, glucose 6-phosphate and dihydroxyacetonephosphate, between cytosol and plastid became redundant by the introduction of the artificial cytosolic-plastidic compartment (See chapter 6.3 for more details). So far, a transporter for acetyl-CoA has not been discovered and was therefore not incorporated (Schwender et al., 2006). CO_2 was assumed to freely diffuse within the cell.

Anabolic pathways for biomass synthesis. The compositional characteristics of rice seedlings were determined experimentally for fatty acids, organic acids, proteinogenic amino acids, free sugars and starch (Appendix 10.1 and Fig. 4-5 A), whereas values for nucleotide and pigment content were taken from the literature (Murray & Thompson, 1980; Panda & Sarkar, 2013; Sumiyoshi et al., 2013; Suzuki et al., 2001; Yu & Zhang, 2013). The cell wall composition was taken from literature (Sumiyoshi et al., 2013) and its fraction was inferred from the measured carbon content of 40.8 %. The overall composition of biomass was not similar to that of *A. thaliana* leaves. In rice seedlings, more than 40 % of the assimilated carbon is captured in soluble and insoluble protein, which is rather high. However, a correspondingly high value has previously been reported for rice seedlings in the literature (Hashimoto et al., 1989). Also, it has been shown that developing leaves require many proteins, of which RuBisCO displays the highest expression (Hashimoto & Komatsu, 2007). Furthermore could the lower contribution of cell wall components to the overall carbon balance be attributed to the immaturity of the tissue, as the cell wall fraction is known to increase during leaf development (Cosgrove, 2005). Also the ash-content is rather high (14.5 %), although it seems to match literature values of about 13-15 % quite well (Misra et al., 2006). Anabolic pathways, providing the building blocks for growth, are spread across the different compartments, with most carbon originating from the cytosolic-plastidic compartment (Fig. 4-5 B). For instance, the pathways of amino acid biosynthesis were responsible for the carbon sink from the cytosolic-plastidic compartment, except for the mitochondrial production of glutamate, proline and arginine and the plastid-specific synthesis of aromatic amino acids (MetaCrop, release 2.0; Reyes-Prieto & Moustafa, 2012). Also free sugar metabolism was located in the cytosolic-plastidic compartment (OryzaCyc, version 2.0), whereas fatty acid synthesis originated from acetyl-CoA in the plastid (Rawsthorne, 2002). Starch was accumulated in the plastid (Orzechowski, 2008).

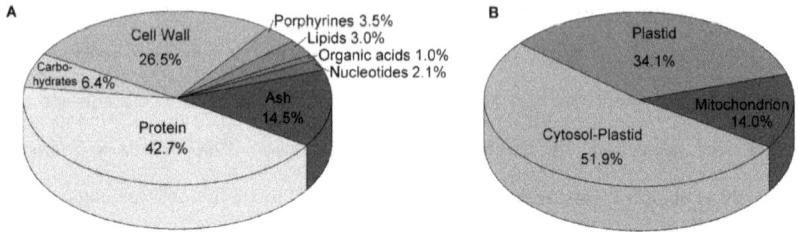

Figure 4-5: Metabolic composition of *O. sativa* seedling shoots

(A) Carbon content-normalized biomass composition of *O. sativa* seedling shoots, based on the information given in Appendix 10.1 (expressed in %) (B) Compartment-specific carbon demand for growth of *O. sativa* seedling shoots (expressed in %), which was calculated by combining the stoichiometric precursor demand (Appendix 10.2) of each biomass constituent with its relative composition (Appendix 10.1).

Export to the root. Seedling shoots additionally export photosynthetic assimilates to the root. Through comprehensive GC-irMS analysis of root tissue in the course of a $^{13}CO_2$ labeling study with 15 day old rice seedlings under ambient conditions, the metabolites responsible for carbon transport were identified (Fig. 4-6). This time-dependent incorporation study revealed that the majority of exported carbon to the root, was in the form of sucrose. The total label incorporation into root sucrose was very high as compared to other root sugars and amino acids under investigation. Although also alanine, glycine and serine showed a rather high ^{13}C enrichment (Appendix 10.6), carbon transport into these pools was negligible, as their concentrations were very low. Therefore, only sucrose export to the root was considered in the network.

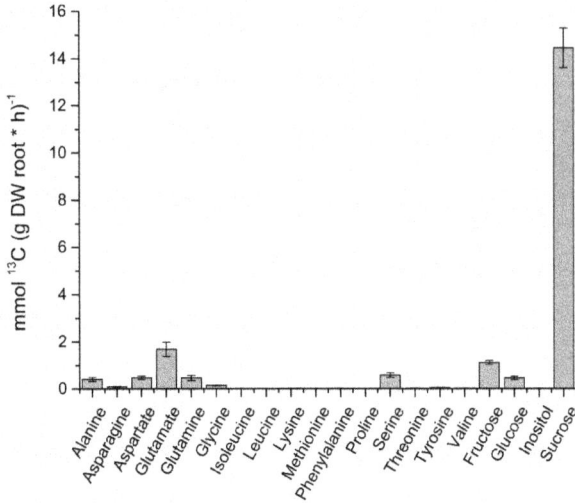

Figure 4-6: Translocation of carbon to the root

The total carbon percentage of different sugars and amino acids translocating into *O. sativa* roots, as determined by the respective MID (by GC-irMS analysis) and pool size measurement. Root samples were collected at 300, 600 and 1800 s during a $^{13}CO_2$ labeling study with 15 day old rice seedlings under ambient conditions.

5.   *In silico* stoichiometric network analysis of *A. thaliana*

In the previous chapter, a well-defined large-scale metabolic network for the major model organism in plant biology and biotechnology, *Arabidopsis thaliana*, was established, which now allows *in silico* investigation of different physiological scenarios of primary plant metabolism. A systematic analysis of this metabolic network, utilizing the framework of elementary flux modes, investigates metabolic capabilities and predicts highly relevant properties on the systems level: optimum pathway use for maximum growth and metabolite overproduction, and flux re-arrangement in response to environmental perturbation. In addition, it allows the integration of computational predictions with experimental fluxome and transcriptome data towards higher-level understanding of plant metabolism. Furthermore, strategies are explored to develop superior plant lines by *in silico* target-prediction. The successful application of predictive modeling in *Arabidopsis thaliana* can bring systems-biological interpretation of plant systems to the next level and, thus, open doors for the metabolic engineering of important crops and for sustainable production of therapeutics, renewable chemicals and biofuels.

5.1.   Analysis of diurnal physiology by elementary flux modes

Plants are subjected to changing environmental conditions, most prominently the light-dark shift. During the day, light is available as a copious source of energy, allowing photosynthetic carbon assimilation, whereas during the night the breakdown of acquired starch delivers the necessary energetic power (Smith & Stitt, 2007). These two fundamental growth states, chemoheterotrophy and photoautotrophy, were now studied for *A. thaliana* using elementary flux mode analysis.

5.1.1. Theoretical analysis of growth physiology of A. thaliana

Plant metabolism displays highly efficient carbon assimilation and conversion. The elementary flux solutions, which span the entire space of feasible flux distributions, were calculated both for autotrophic metabolism in the light, using CO_2 as sole carbon source, and for heterotrophic metabolism on internal starch in the dark, respectively. The solution space consisted of 1.2 million autotrophic modes and 5.7 million heterotrophic modes (Table 5-1). The theoretical maximum growth yields were 28.6 and 27.4 (g biomass) (C-mol substrate)$^{-1}$ for the chemoheterotrophic and for the photoautotrophic scenario, respectively (Table 5-2). Assuming equal contributions of day and night metabolism, the resulting mean value of 28.1 (g biomass) (C-mol substrate)$^{-1}$ rather matches with experimental values for wild type A. thaliana leaves of about 25.2 (g biomass) (C-mol substrate)$^{-1}$ (Table 5-2) (Sulpice et al., 2014). For prolonged illumination phases, the simulated values further approach the measured rates (Appendix 10.7). The maximum theoretical biomass formation accounted for a high carbon efficiency, i.e. 86.8 and 90.6 % of the assimilated carbon was incorporated into biomass. This can be taken as a first indication that plant metabolism can closely approach theoretical optimum performance with regard to stoichiometric capacity. When the light influx was omitted from the model, no modes resulted for autotrophic biomass production, deducing that heterotrophy is the only feasible phenotype in the dark.

A narrow range of absorbed photons supports optimal plant growth. The computed set of elementary modes was now used to investigate the impact of the light influx on biomass production. First, energetically inefficient modes that included cycling of resources, were eliminated. Although stoichiometrically possible, these modes represented futile cycling of carbon across the internal membranes to dissipate energy (Appendix 10.11). They are an emergent property from the compartmentation of metabolic pathways into intracellular organelles. As such massive futile cycling is energetically inefficient, they are considered both evolutionary and physiologically implausible. Furthermore, as such high reaction rates require additional synthesis of large amounts of protein, for which the synthesis cost was not accounted

for in the biochemical model, it was opted to eliminate these modes for subsequent analyses.

Table 5-1: Outline of fundamental physiological states in plant leaves and their accompanying EFMs.

	Autotrophy		Heterotrophy	
	Light	Dark	Light	Dark
CO_2 Uptake	+	+	-	-
Starch Degradation	-	-	+	+
Light influx	+	-	+	-
Number of EFMs	1 199 402	11	11 296 607	5 653 544

Table 5-2: Maximal theoretical biomass production of *Arabidopsis* leaves as predicted by the metabolic

model, both photoautotrophically in the light and chemoheterotrophically in the dark, are compared to the

experimentally determined growth yield of *Arabidopsis* rosettes. The latter can be found in the far right

column and was calculated based on the experimental work of Sulpice et al. (2014). See Appendix 10.7

for more information on the experimentally determined biomass yield.

	Photoautotrophy model	Chemoheterotrophy model	Whole plant experimental
Substrate	CO_2	C_{12}-starch subunit	CO_2
Substrate Uptake [mmol substrate]	1	1	1
Biomass Production [mg Biomass]	28.6	329	25.2
Biomass Yield [g / C-mol substrate]	28.6	27.4	25.2
Carbon Efficiency [%]	90.6	86.8	79.9

For the resulting 848 629 individual flux modes, the biomass was mapped against the corresponding acquired light influx (Fig. 5-1). The biomass production was highest in a defined range between 8.6 and 10.8 (mol photons) (mol CO_2)$^{-1}$. This corresponded to a quantum yield between 0.09 - 0.12 (mol CO_2) (mol photons)$^{-1}$, supporting optimal photosynthesis. Experimental growth for *Arabidopsis* rosettes closely resemble the predicted theoretical maximum and the predicted optimum range of photon absorbance rather exactly matched with the range observed *in vivo* (Ehleringer & Bjorkman, 1977; Ehleringer & Pearcy, 1983; Long et al., 1993; Osborne & Garrett, 1983; Skillman, 2008).

5.1.2. Systematic analysis of elementary mode solution space unravels light stress response of *Arabidopsis thaliana* leaves

Optimal growth only resulted for a narrow range of photon absorbance as both higher and lower light intensities reduced growth efficiency (Fig. 5-1). Possible molecular reasons for this phenomenon can be directly extracted from the flux mode solution space. The direct product of photo-reduction, is NADPH (Fig. 5-1 B). Under high light intensity, its production soon exceeds the need of the anaplerotic reactions. A physiologically feasible way to cope with such NADPH excess is an increased flux through the photorespiratory pathway, as photorespiration consumes NADPH. Interestingly, a clear relationship between the flux through the photorespiratory system (glycerate kinase), phosphoribulokinase, and the quantum requirement under high light intensity suggests that indeed *Arabidopsis* leaves may deal with photoreductory stress by increasing photorespiration (Fig. 5-1 D-E). A direct consequence of such an up-regulation is reduced growth, because photorespiration is associated with a net loss of carbon as CO_2 (Ma et al., 2014; Poolman et al., 2013). Recently, an activated photorespiration has been experimentally recognized in plants as an important light stress response to dissipate excess reducing equivalents and energy (Voss et al., 2013). Alternatively, NADPH excess could be also handled by an increased consumption of ATP, e.g. through futile cycles. From a metabolic viewpoint, this would involve a conversion of excess NADPH into NADH by a transhydrogenase-like reaction, and fueling the phosphorylation of ADP into ATP. In this way, growth could be maintained at its optimal rate, however in an energetically inefficient way.

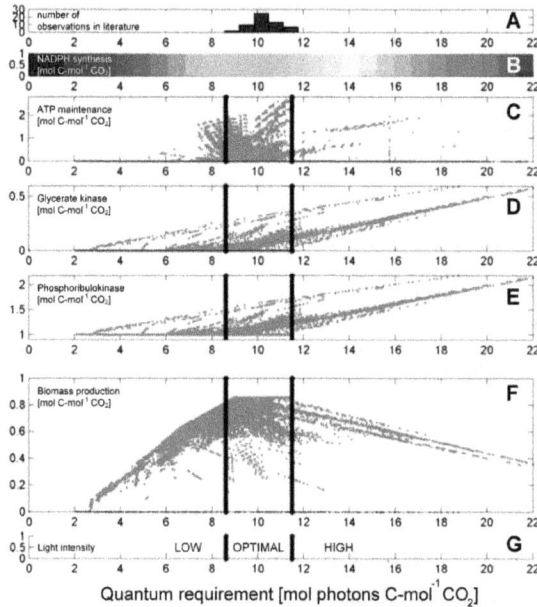

Figure 5-1: The quantum requirement for growth.

(A) Histogram of the quantum yield values for C_3 plants under ambient atmospheric conditions were selected from Skillman (2008) who reviewed the *in vivo* range of quantum yield in several C_3 and C_4 plant lineages. Not all harvested quanta are converted into chemical energy as some are lost through absorption by pigments which are unable to contribute their excitation energy to photosynthesis. Therefore, the nominated range was corrected assuming 47 % of photons are outside the photosynthetically active range (Hall & Rao, 1999). (B) The linear relationship between modeled NADPH synthesis and quantum requirement is visualized as a colorscale between blue (low) and red (high). (C) Modeled ATP maintenance production associated with the observed light influx. (D) Modeled glycerate kinase and (E) phosphopentokinase activity as a function of photon influx. (F) Observed biomass formation in relation to the quantum requirement. (G) Subdivision of the data into three light regimes. Panel C-G: Each data point represents an individual flux mode.

5.1.3. Flux rearrangement between optimum day and night metabolism requires reversible translocation of carbohydrates across the plastidial and mitochondrial membranes, as well as strong regulation of plastidial metabolism

Next, day-time and night-time metabolism were evaluated on the level of intracellular fluxes. The intracellular flux distributions, which resemble optimum growth under both conditions, revealed large differences (Fig. 5-2 and 5-3). Most interestingly, our data show an on/off shift for five previously postulated diurnal switches, i.e. for ribulose 1,5-bisphosphate carboxylase, phosphoribulokinase, fructose 1,6-bisphosphate phosphatase, NADP-dependent glyceraldehyde-3-phosphate dehydrogenase and sedoheptulose-1,7-bisphosphate phosphatase (Taiz & Zeiger, 2006). In addition, many more changes were observed. A reversed flux direction was observed for no less than 21 reactions and 89 reactions showed substantial changes in flux value. From the latter, 64 fluxes were significantly decreased for the dark metabolism as compared to the light metabolism, whereas 25 reactions showed a notable increase. The largest differences were found for reactions in the plastid, the energy metabolism and for specific transporters, which is essentially caused by the shift from starch degradation in the dark to CO_2 assimilation in the light. In contrast, especially the mitochondrial and peroxisomal reactions remained rather unchanged. Shortly, dark metabolism required degradation of starch into maltose, which was subsequently transported across the plastidial membrane. In the cytosol, maltose was hydrolyzed to two glucose molecules, followed by phosphorylation to glucose 6-phosphate. The sugar phosphate was then channeled into the cytosolic and the plastidic EMP pathways. The glucose 6-phosphate translocator mediated the distribution of carbon between the different compartments. In this way, both malate and pyruvate were generated through malate dehydrogenase and glycolysis, respectively, and subsequently imported in the mitochondrion to fuel the TCA cycle. Autotrophic metabolism assimilated atmospheric CO_2 via the CBB cycle in the plastid. This involved a high flux through the CBB cycle and the non-oxidative PP pathway, as both routes are strongly entwined. Mainly, the triose-phosphate

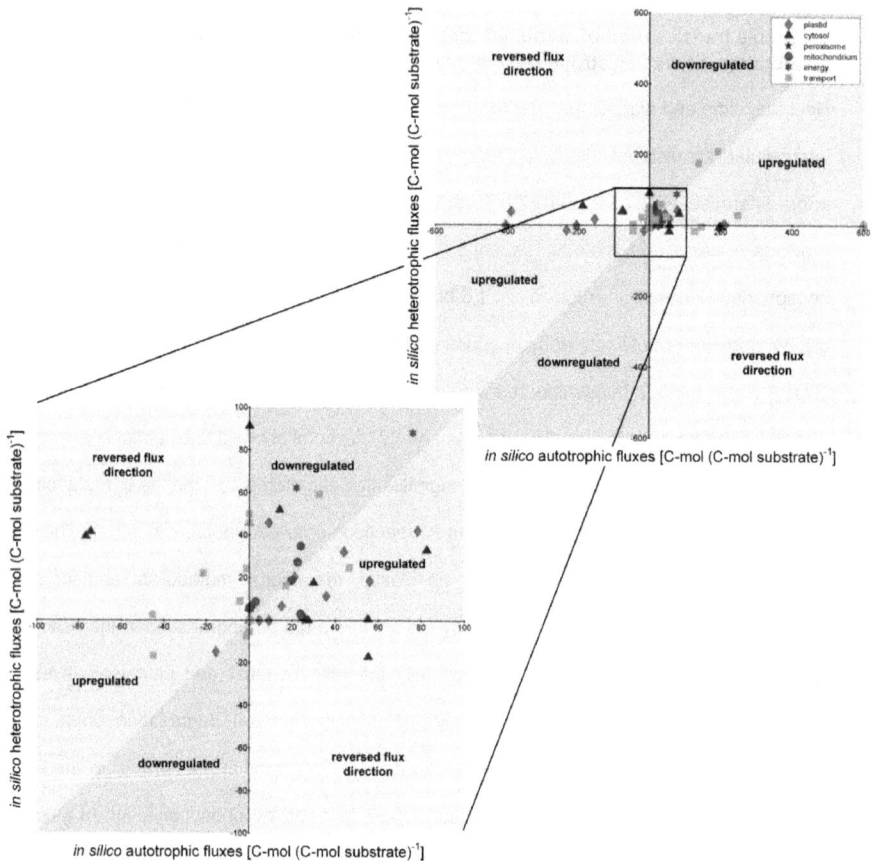

Figure 5-2: Autotrophic versus heterotrophic fluxes

Fluxomic phenotype of optimal autotrophic plant growth as compared to optimal heterotrophic growth. The second and fourth quadrant are affiliated with a reversed flux direction, whereas the first and third quadrant are further subdivided into reactions that were up- or downregulated in autotrophic growth. In this context, 'optimal' refers to the average flux value of the top 1 % biomass producing modes. All fluxes are normalized to the substrate uptake flux and are expressed in C-mol C-mol^{-1} substrate.

translocators supplied carbon-building blocks to the cytosol, whereby the glucose 6-phosphate translocator recycled carbon back into the plastid. Mainly malate was built from initial carbon conversions and imported into the mitochondrion. Here, malic enzyme provided the TCA cycle with sufficient pyruvate. In this regard, the translocation of pyruvate between the different compartments was one prominent example of a fully reversed flux, as described above. To conclude, in the light, especially the plastidic metabolism was highly active and the triosephosphate-carriers transported carbon out of the plastid. Mainly malate was imported into the mitochondrion, whereas in the dark both malate and pyruvate were channeled across the mitochondrial membrane and the cytosol showed slightly more activity than the plastid. The main carbon export from the plastid occurred in the form of maltose.

Figure 5-3 (next page): Fluxes of the day and night metabolism

Maximal biomass producing fluxes in the photoautotrophic (left) and chemoheterotrophic (right) case. The value on the arrow and the thickness of the arrow represent the average flux value of the top 1 % biomass producing modes. All fluxes are normalized to the substrate uptake flux and are given in mol mol^{-1} substrate. To enable comparison, the autotrophic flux values were normalized to 600 mol of CO_2 uptake and the heterotrophic modes to 50 mol of starch degradation, as starch is represented by C_{12}-dimers. For clarity the arrow thicknesses are normalized to the uptake flux thickness. The arrows pointing to the *Arabidopsis* rosettes visualize the amount of building blocks needed for biomass synthesis.

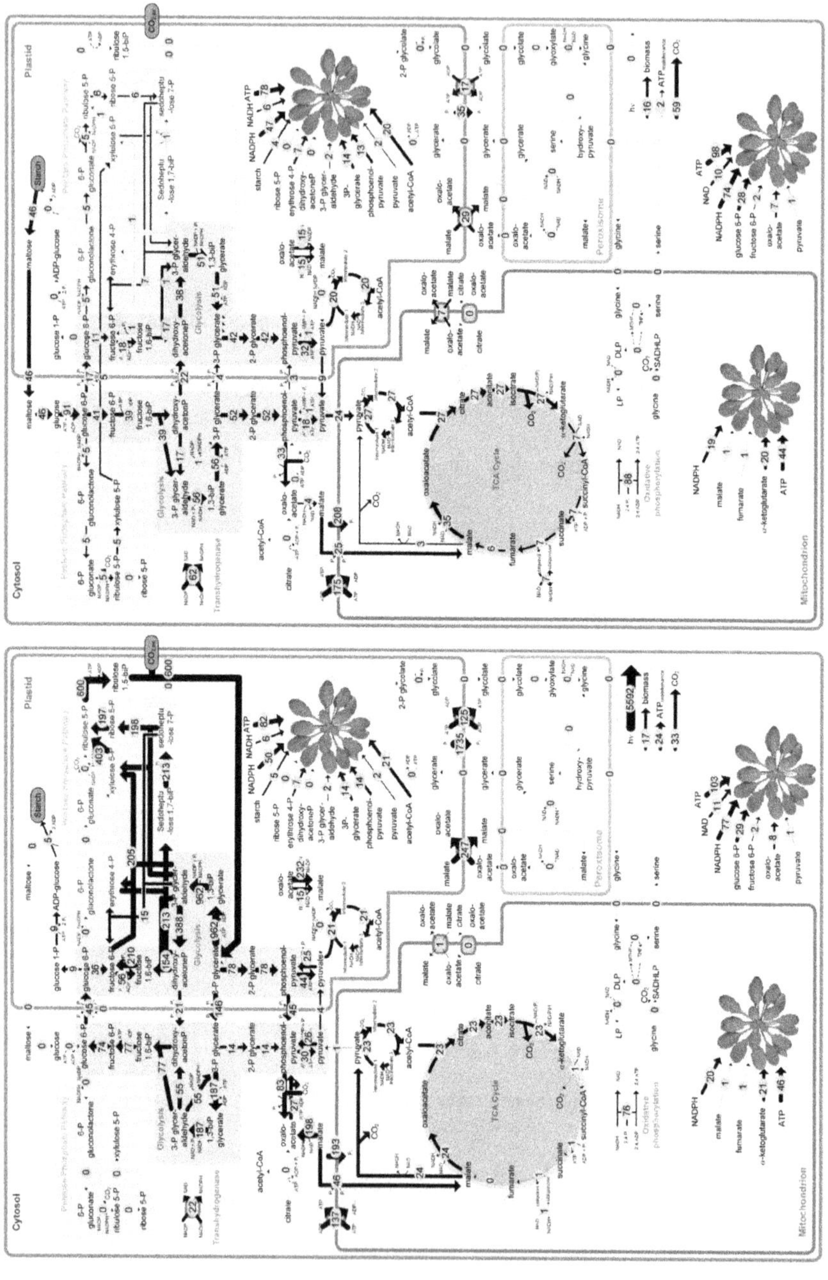

5.1.4. Simulation of light and dark metabolism quantifies differences in energy and redox supply

As shown, the metabolism in the dark recruited all compartments to sustain growth, whereas the plastid was the dominant compartment in the light (Fig. 5-3). Next, it was investigated which organelles support the supply of the necessary energy and reductive power for the growing leaves (Fig. 5-4). Illuminated leaves largely delivered NADPH and ATP through the photosynthetic light reactions in the plastid (Fig. 5-4 A), whereas the NADH stemmed mostly from cytosolic malate dehdrogenase, i.e. is exported as NADH from the plastid (Cheung et al., 2013). During the dark, ATP was produced predominantly via oxidative phosphorylation in the mitochondrion (60 %), and reduced NADPH was generated by all compartments (Fig. 5-4 B). Secondly, the formation and assimilation of carbon dioxide was different between the two physiological states (Fig. 5-4 C). Generally, the total net flux of CO_2 production was higher for heterotrophic metabolism (10 %), as compared to autotrophic metabolism (5 %), whereby overall, both values were rather low. It was interesting to note that also the origin of the released CO_2 differed. Under both conditions, the carbon loss by photorespiration was negligible, whereas the largest contributors to CO_2 release were the TCA cycle, as well as pyruvate and malate dehydrogenases (Fig. 5-4 D). In the dark, additional carbon was released during the generation of NADPH in the oxidative PP pathway, whereas during illumination, previously assimilated CO_2 was lost via PEP carboxykinase.

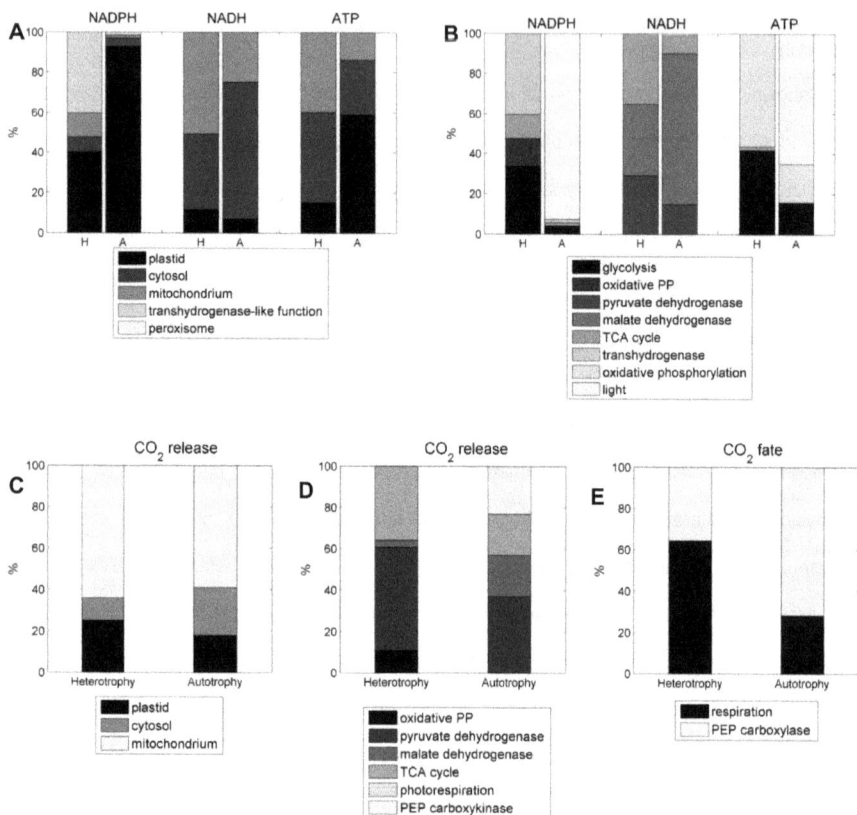

Figure 5-4: Metabolic properties of light and dark metabolism

Comparison of metabolic properties for optimal autotrophic and heterotrophic fluxome as calculated by elementary flux analysis. In this context, 'optimal' refers to the average flux value of the top 1 % biomass producing modes. (A-B) Specifications about the redox and energy production expressed per total redox and ATP-demand. (A) Pathway-based subdivision and (B) compartment-based visualization. (C-E) Details about the origin and fate of CO_2 as percentage of the total CO_2 released. Here, the fluxes were subdivided into its different (C) compartments and (D) pathways.

5.1.5. Optimum anabolic pathway use in *A. thaliana* differs between day and night

The striking differences in the localization of energy and reducing power supply among the cellular compartments now suggested to also inspect the anabolic metabolism. Interestingly, the biosynthetic origin of many of the twelve anabolic precursors strongly depended on the physiological growth mode, i.e. light and carbon acquisition (Fig. 5-5). As example, the majority of autotrophic 3-phosphoglycerate was supplied via the CBB cycle, whereas, in the dark, the precursor exclusively stemmed from the EMP pathway. In addition, ribose 5-phosphate was produced through the non-oxidative PP pathway under illumination, whereas, under heterotrophic conditions, it was mainly generated by the oxidative PP pathway. For other precursors, such as α-ketoglutaric acid, the metabolic network recruited isoenzymes, which differed in cofactor use, depending on the illumination conditions.

Clearly, these simulations indicate that a reversible transhydrogenase-like function is crucial to rearrange the metabolism from day to night (Fig. 5-4 A). It is interesting to note that dark and light metabolism involves a rather diverse set of anabolic pathways for optimum precursor supply (Fig. 5-5). This also involves isoenzymes with different prevalence for NADPH and NADH, which vary between the two growth regimes. So far, their role is still unclear, however functional diversification (Costenoble et al., 2011) and increased metabolic robustness (Blank et al., 2005) have been postulated as possible purposes of isoenzymes. Based on our simulations, the ubiquitous presence of isoenzymes with different cofactor usage in plants seems key for flexible handling of specific anabolic demands of day and night metabolism.

Figure 5-5 (next page): Pathways involved in the allocation of anabolic precursors

Percentaged information about the pathways involved in allocating the necessary anabolic precursors under optimal autotrophic and heterotrophic conditions. In this context, 'optimal' refers to the average flux value of the top 1 % biomass producing modes. Abbreviations: 3-phosphoglycerate (3PG), acetyl-CoA (AcCoA), α-ketoglutaric acid (AKG), erythrose 4-phosphate (E4P), fructose 6-phosphate (F6P), glucose 6-phosphate (G6P), glyceraldehyde 3-phosphate (GAP), oxaloacetate (OAA), phosphoenolpyruvate (PEP), pyruvate (PYR), ribose 5-phosphate (R5P), succinyl-CoA (SucCoA).

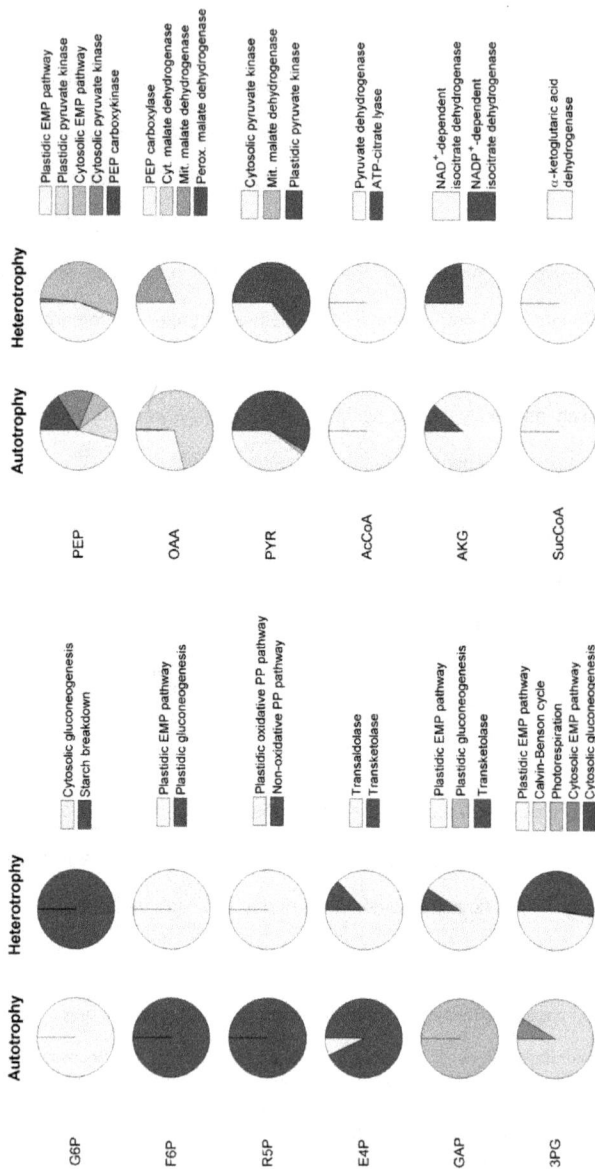

5.2.   Data-integrated analysis of elementary flux mode solution space

The constructed metabolic network enables the adequate description of fundamental physiological traits such as growth and photosynthetic efficiency, which is a good indication of its quality and validity (Fig. 5-1 and Table 5-2). The structural analysis of network topology already showed that the diurnal day-night shift of *Arabidopsis* requires substantial re-arrangement of pathway flux between compartments, involving redox and energy metabolism, as well as anabolic synthesis. Now, higher-level understanding of plant metabolism can be gained by integrating these computational network predictions with experimental fluxome and transcriptome data.

5.2.1. The intracellular pathways of *Arabidopsis* rosettes operate near their optimum flux distributions

The integration of *in silico* and *in vivo* fluxes allows evaluation of the observed cellular physiology within the overall feasible flux space, as proven valuable for different microorganisms (Becker et al., 2011; Driouch et al., 2012). Fortunately, recent ^{13}C flux analysis data from *Arabidopsis* rosettes (Szecowka et al., 2013) and from plant cell cultures (Masakapalli et al., 2014), have provided an excellent opportunity to conduct such an integration for the first time in plant cells. For this purpose, the determined flux values were now mapped with the predicted fluxes for optimum growth (Fig. 5-6). This provided a striking agreement between the *in vivo* and optimum *in silico* fluxome. Considering the similarity of measured and predicted biomass and quantum yield, we conclude that illuminated thale cress is represented adequately by the proposed metabolic model, and that its leaves operate very close to their maximal potential with regard to stoichiometric capacity. It should be noted that this may not hold for suboptimal growth conditions, such as high light acclimation (Ma et al., 2014), as stress conditions seem to induce a metabolic burden.

In contrast, the metabolism of heterotrophic cell cultures (Masakapalli et al., 2014) deviated from

the optimum dark growth mode (Fig. 5-6). This is particularly true for the TCA cycle and the EMP

pathway. Obviously, our leaf model does not explain fluxes in heterotrophic cell cultures

adequately. This highlights the need for physiologically relevant models, as heterotrophic cell

cultures display a fundamentally different physiology and are known to respond differently to

stress on a metabolic level than live plants (Karmakar et al., 2009; Lehmann et al., 2009).

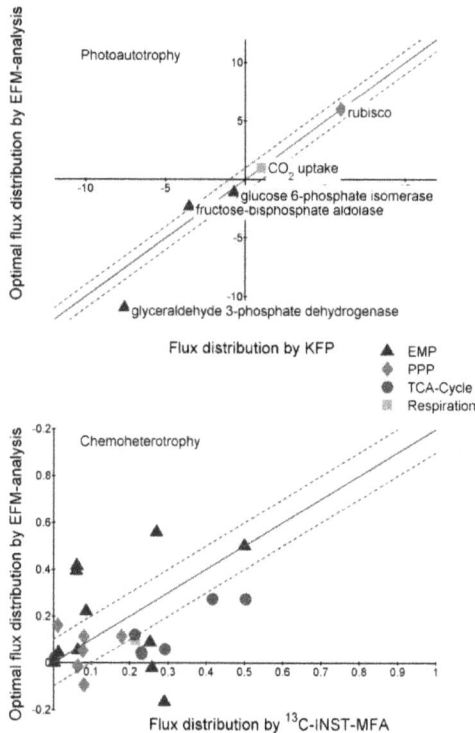

Figure 5-6: *In silico* versus *in vivo* metabolic fluxes

The metabolic phenotype of an illuminated *Arabidopsis* rosette (A) and the heterotrophic phenotype of an *Arabidopsis* cell culture (B), determined by ^{13}C-MFA (Ma et al., 2014) or KFP (Masakapalli et al., 2014), are compared to the respective flux mode yielding optimal biomass. In this context, 'optimal' refers to the average flux value of the top 1 % biomass producing modes. All fluxes are normalized to the substrate uptake flux and are expressed in C-mol (6 C-mol substrate)[-1].

5.2.2. Day-night flux re-arrangements are superimposed by selected transcriptional changes

As shown, the optimum metabolism of *Arabidopsis* recruits a different set of pathways during day and night: diurnal cycling is linked to substantial redistribution of flux. It was now interesting to see, how regulatory circuits of the plant superimpose the obvious fine-adjustments on the level of the metabolic network, i.e. to which extent transcriptional and post-transcriptional control is involved, and which genes are part of the metabolic re-arrangement. Our simulation data now enable integration of fluxes and the expression of encoding genes. Such integration of transcriptomic knowledge with observed flux changes offers a systems level view of the day-night shift (Fig. 5-7). Reactions, such as the ones catalyzed by mitochondrial and plastidic pyruvate dehydrogenase and citrate synthase are stable in either fluxome or transcriptome. Others display clear diurnal trends, as for instance cytosolic and plastidic pyruvate kinase, phosphoribulokinase, fructose 1,6-bisphosphate phosphatase and many others. Even for those genes that appear to have unaltered expression levels, although fluxomic changes are observed, isogenes exist that are still to be tested, including reactions of the oxidative PP pathway, fumarase or hexose isomerase. Diurnal changes in both transcripts and fluxes were observed for the previously identified on/off switches, i.e. for ribulose 1,5-bisphosphate carboxylase, phosphoribulokinase, fructose 1,6-bisphosphate phosphatase, NADP-dependent glyceraldehyde-3-phosphate dehydrogenase and sedoheptulose-1,7-bisphosphate phosphatase (Taiz & Zeiger, 2006). In addition, the triose phosphate translocator between plastid and cytosol is known to follow a diurnal path for carbon export (Taiz & Zeiger, 2006). This supports the predicted high export flux for 3-phosphoglycerate, phosphoenolpyruvate and dihydroxyacetone phosphate during the light period. Additionally, light/dark modulation of pyruvate-orthophosphate (Pi) dikinase through phosphorylation in C_3 leaves (Chastain et al., 2002) is confirmed by an off switch during the night. Overall, systems level correlation occurs plentiful and seems well distributed over the entire central carbon metabolism, reflecting the necessity of readjusting the core metabolism during day/night-transition.

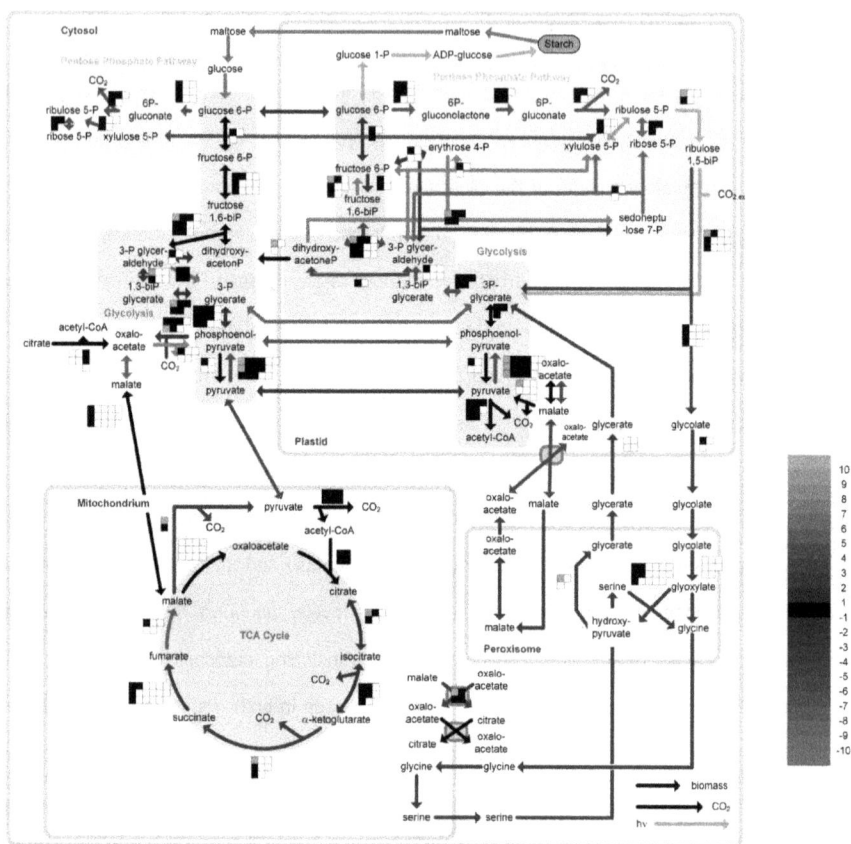

Figure 5-7: Integration of fluxes and transcriptome

Color-coded visualization of amplitude in flux change between optimal dark and light metabolism ($\log_2$ light/dark) as calculated by the proposed model. In this context, 'optimal' refers to the average flux value of the top 1 % biomass producing modes. Green arrows represent positive amplitudes, whereas blue arrows stand for negative changes. Transcriptome data from Bläsing et al. (2005), Chastain et al. (2002) and Taiz and Zeiger (2006) are captured in the accompanying boxes. Here genes with no significant change in the amplitude are depicted in black and genes with large absolute amplitude are shown in green. White boxes represent untested genes.

A more detailed view on selected transcriptional changes is obtained by overlaying the time resolved expression of metabolic genes from *Arabidopsis* grown with day/night cycles (Bläsing et al., 2005) with the corresponding flux changes of the encoded reactions, as predicted from the simulations (Fig. 5-8). In a number of cases, transcription-based and flux-based changes exhibit a close connection, which indicates that the plant strongly recruits regulatory mechanisms to drive core metabolism. Interestingly, the genes that showed direct correlation between transcript and flux, were located around three main controlling points. Firstly, the fluxes between fructose 6-phosphate and triose-phosphate appeared highly regulated as both fructose 1,6-bisphosphatase and fructose bisphosphate aldolase showed high correlation. Secondly, many enzymes involving malate and pyruvate had a diurnal pattern, indicating malate and/or pyruvate might be a second controlling point in metabolism. Both the fructose 1,6-bisphosphatase and pyruvate kinase controlling points have previously been postulated as diurnal regulators of metabolism (Steer, 1974). In addition, the complex regulation pattern around the malate node is reflected in the local high network complexity. Finally, genes associated with the TCA cycle seem superimposed by transcriptional regulation.

Besides identifying clusters of potential transcriptional control, this type of analysis can be used to identify, which isoenzymes are likely linked to flux changes. For example, for three pyruvate kinase-associated genes (At5g08570, At5g56350 and At3g49160) a diurnal expression pattern was identified, however only the former two are linked to the observed flux changes. This indicates that At3g49160 likely has a functionally different purpose from mediating the pyruvate phosphorylation in central carbon metabolism. For instance, pyruvate kinase could also catalyze several metabolic conversion in ribonucleotide and nucleotide biosynthesis. As isoenzymes might have distinct functions (Costenoble et al., 2011), correlation of flux and transcript could pose a valuable method for identification of isoenzymatic function and might provide first evidence on their biochemical function. This is particularly useful in cases, where a high number of isoenzymes potentially catalyze a certain reaction.

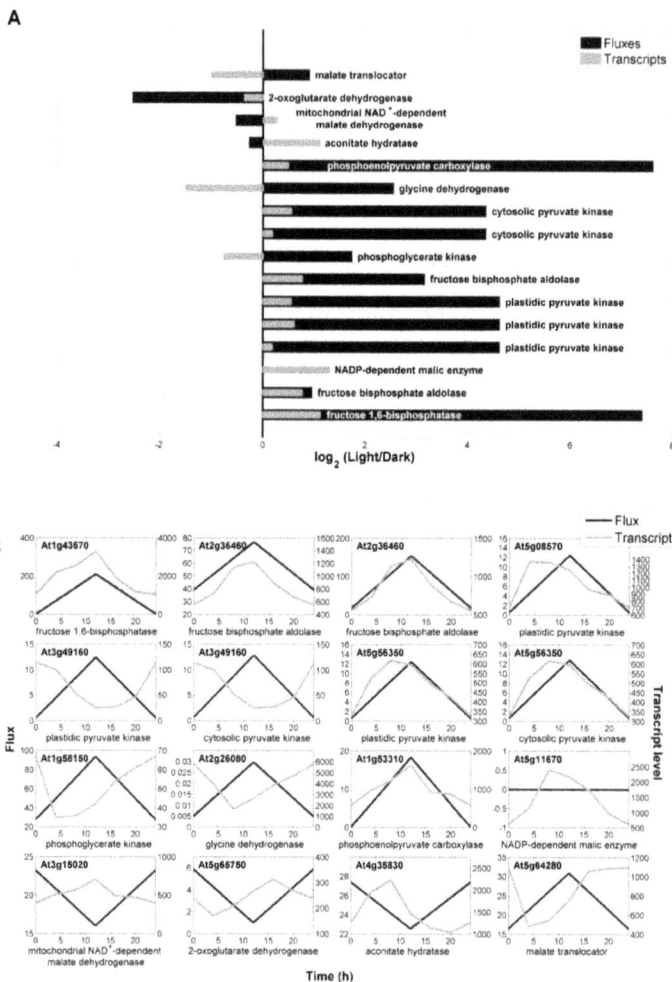

Figure 5-8: Comparison between transcriptome and calculated optimal fluxes

(A) Comparison of the amplitude in diurnally expressed genes from Bläsing et al. (2005) and the amplitude in flux change calculated by the model. (B) Gene expression profiles for several genes from the central metabolism, as determined by Bläsing et al. (2005), were overlaid with their respective flux change from our simulations. Optimal photoautotrophic and chemoheterotrophic flux modes were selected to be representative of day and night time metabolism, respectively. In this context, 'optimal' refers to the average flux value of the top 1 % biomass producing modes. Gene expression data are displayed with their expression level, whereas the fluxome is given in C-mol C-mol[-1] substrate. A 12 h light/12 h dark diurnal cycle was adopted.

Furthermore, flux-transcript correlation can assist in the confirmation of genes with putative function. Putative genes that do not show significant linkage to flux change, possibly do not control flux on a transcriptional level (At3g15020, At2g26080, At1g58150, At3g49160, At5g11670). However, those that do correlate, might present candidates involved in fructose bisphosphate aldolase (At2g36460), fructose 1,6-bisphosphatase (At1g43670) and phosphoenolpyruvate carboxylase (At1g53310). One should keep in mind for these interpretations, that not in all cases transcript changes will immediately lead to protein change, because translation in plants can be damped or delayed (Baerenfaller et al., 2012).

Taken together, the metabolic simulations provide detailed molecular insights into plant functioning. It seems that *Arabidopsis* can operate close to its theoretical pathway optimum and that this is mediated by a fine-adjustment of metabolic flux, strongly under transcriptional control. In this light, the present work is one of the very few examples so far, which link *in vivo* with *in silico* flux data to a higher-level understanding (Driouch et al., 2012; Moisset et al., 2012; van Duuren et al., 2013).

5.3.   Flux-homeostasis and biotechnological impact of plasticity in

photophosphorylation

Stable optimal fluxes under photosynthetic plasticity. It is generally accepted that both cyclic and non-cyclic electron flow contribute to the *in vivo* photosynthetic light reactions (Allen, 2003; Cruz et al., 2005; Kramer et al., 2004), however it is unclear in which ratio cyclic and non-cyclic electron flow co-occur. It is hypothesized that the observed flexibility allows modulation of the NADPH:ATP ratio to match the demands of metabolism under changing environmental conditions (Cruz et al., 2005; Kramer et al., 2004) as non-cyclic electron flow allows for both NADPH and ATP synthesis, whereas cyclic electron flow solely generates ATP. It was now interesting to investigate the interplay between the metabolic pathways in meeting different metabolic ratios for NADPH and ATP. For this purpose, autotrophic flux distributions were calculated for 15 scenarios with increasing contribution of cyclic electron flow to the overall electron flow. Ratios between non-cyclic and cyclic electron flow of 14:0, 13:1, 12:2, 11:3, 10:4, 9:5, 8:6, 7:7, 6:8, 5:9, 4:10, 3:11, 2:12, 1:13 and 0:14 were considered. For all ratios, except for 100 % electron flow, the same theoretical maximum for biomass formation was found. Omission of non-cyclic electron flow did not allow for any biomass formation. Furthermore, three general trends could be observed when plotting the absolute flux values against the ratio between cyclic and non-cyclic electron flow (Fig. 5-9). Either optimal flux values were not influenced by the type of phosphorylation (e.g. ribose 5-phosphate isomerase) or an exponential increase at low or high cyclic contribution was observed (e.g. mitochondrial phosphate import and glyceraldehydes 3-phosphate dehydrogenase, respectively). In the range between 20 and 70 % cyclic electron flow contribution all fluxes remained rather unchanged. This robustness can be taken as an indication that plant metabolism is highly resilient, easily capable of adjusting to different NADPH:ATP ratios.

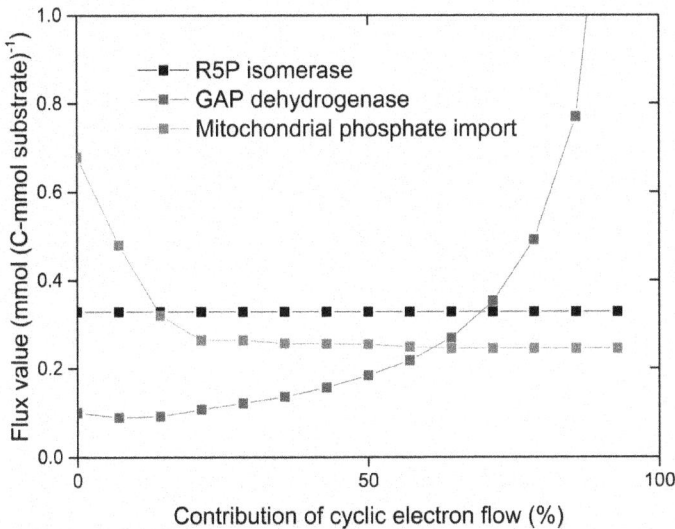

Contribution of cyclic electron flow (%)

Figure 5-9: Flux changes with changing ratio between cyclic and non-cyclic electron flow

Trends in flux changes of optimal fluxes as calculated for increasing contribution of cyclic electron flow to photophosphorylation. In this context, 'optimal' refers to the average flux value of the top 1 % biomass producing modes. All fluxes are normalized to the substrate uptake flux and are expressed in mmol C-mol⁻¹ substrate.

Metabolic adaptation to changing light environment. Plants adapt to changes in light intensity through photoprotection and optimization of energy conversion (Cruz et al., 2005). In this way, the output ratio of ATP:NADPH can be influenced significantly by the light environment, which has been investigated extensively (Allen, 2003; Cruz et al., 2005; Kramer et al., 2004). However, how cellular metabolism copes with the changed ATP:NADPH supply on the level of intracellular fluxes is investigated here for the first time. The observed stability of biomass formation, with only few distinct flux changes, across a wide range of ATP:NADPH ratios (Fig. 5-9) indicates a certain robustness of metabolism to environmental changes. Furthermore, the metabolic fluxes that are most influenced by small changes around the assumed *in vivo* ratio of non-cyclic and cyclic electron flow of 12:2, appear to handle an excess of redox power by channeling NADPH through the malate/oxaloacetate shuttle into the mitochondrion, where ATP is produced by oxidative phosphorylation (Appendix 10.11). In addition, plastidic triose-phosphate is increasingly exported to the cytosolic EMP pathway, thus effectively reducing the plastidic ATP requirement. Possibly, homeostasis of flux and adenylate/redox status exists, through modulation of NADPH and ATP dissipation. A homeostasis of the adenylate status during photosynthesis in a fluctuating environment has previously been indicated, however, here, homeostasis was supposedly attributed to changes in pathway usage (Noctor & Foyer, 2000). Here, we observe that the net flux distribution remains unchanged under abruptly changing light environment, however, strong differences occur in substrate cycling. The occurrence of such substrate cycling has previously been identified *in vivo* in plant tissues (Alonso et al., 2005). Likely, the observed metabolic adaptation through futile cycling occurs in addition to changes in pathway use, such as increased photorespiration. This improves the plants' capacity to cope with a constantly changing light environment.

Biotechnological advantage of photosynthetic plasticity. Furthermore, it has recently been proposed that such an excess in redox power could be directed towards light-driven production of biotechnological compounds through cytochrome P450s-mediated reactions (Lassen et al., 2014). When an organism is engineered to produce large amounts of a biotechnologically interesting product, its molecular flux patterns change. These metabolic changes engage a different demand for ATP and NADPH, that needs to be accustomed by the cell. Due to the observed plasticity in non-cyclic and cyclic electron flow, plants are capable to adapt to such modulated energetic requirements. In addition, our simulations now show that such adaptations do not impede growth, granting plants a high potential for the production of biotechnological products, especially for those compounds requiring much redox power. This emphasizes the potential of photosynthetic light reactions in biotechnology.

5.4.   Extending correlation-based target prediction for metabolic engineering

strategies to complex networks

Beyond the natural growth boundary of *A. thaliana*, its potential to grow and accumulate specific trait compounds was explored. For this purpose, elementary flux mode-base *in silico* strain design was performed. In this way, the set of elementary flux modes was analyzed with Flux Design to search genetic targets towards improved performance (Melzer et al., 2009).

5.4.1. Towards high-growth yield varieties

It turned out that the high network complexity of the compartmented model hampered straightforward target identification, similar to what is observed in other eukaryotes (Melzer et al., 2009). Especially, the compartmentation of energy, redox and phosphate metabolism, imposed difficulties. However, formation of biomass was not influenced by the imposed compartmentation (data not shown), such that subsequent analyses could be performed with a slightly adjusted network model without energy, redox and phosphate compartmentation (Appendix 10.3). In addition, the correlation analysis was narrowed down to specific subsets of modes, involving only biomass forming modes, the top 5 % and the top 1 % modes with regard to growth. Subsequently, these subsets were tested for their predictive power (Table 5-3). The best subset, i.e. the subset, which provided the highest number of statistically significant targets, was the top 5 % subset of modes. In total, nine correlation targets were identified. Four of these targets directly correlated to biomass formation, as they uniquely produced a particular biomass precursor, and were also found in other subsets and physiology types. Additionally, mitochondrial pyruvate dehydrogenase, citrate synthase, peroxisomal isocitrate lyase, succinate dehydrogenase and fumarase, all located in and around the mitochondrion, (Fig. 5-10) were predicted as potential growth targets. Taking into account the previously observed high flexibility of the plastidic and energy metabolism and pronounced stability of the mitochondrial fluxes

during the diurnal cycle, the localization of targets for genetic engineering in and around the
mitochondrion seems reasonable.

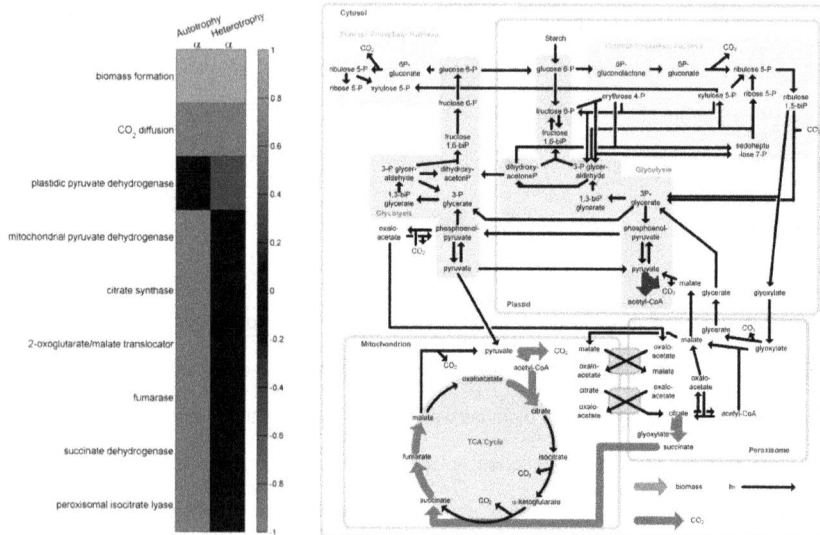

Figure 5-10: Attenuation and overexpression targets as predicted by Flux Design

Attenuation (red) and overexpression (green) targets, encoding specific biochemical conversions as
predicted by Flux Design analysis for the wild type *Arabidopsis* leaf. Left: The color code represents the
fold-increase resp. decrease as predicted by the target potential coefficient α. Right: Visualization of
targets on pathway map. The color code represents the fold-increase resp. decrease as predicted. Only
those targets encoding a metabolic reaction that were accepted by the statistical analysis are displayed.
Transport reactions were excluded from the analysis.

Table 5-3 (next page): Autotrophic and heterotrophic growth targets predicted by Flux Design for defined
subsets of modes. The plus or minus signs preceding the reaction names indicate a positive resp.
negative target potential coefficient. Only metabolic reactions are displayed, transporters were excluded
from the analysis. Also, the subset size is given.

	All modes	Biomass producing modes	Top 5 % biomass producing modes	Top 1 % biomass producing modes
Autotrophy	2 318 951 modes	2 249 604 modes	1 020 487 modes	759 218 modes
Increase	+ in silico biomass transporter	+ in silico biomass transporter	+ in silico biomass transporter	+ in silico biomass transporter
	+ in silico biomass synthesis	+ in silico biomass synthesis	+ in silico biomass synthesis	+ in silico biomass synthesis
	+ plastidic pyruvate dehydrogenase	+ plastidic pyruvate dehydrogenase	+ plastidic pyruvate dehydrogenase	+ plastidic pyruvate dehydrogenase
Decrease	− CO_2 diffusion	− CO_2 diffusion	− CO_2 diffusion	− CO_2 diffusion
	− mitochondrial pyruvate dehydrogenase	− mitochondrial pyruvate dehydrogenase	− mitochondrial pyruvate dehydrogenase	
	− citrate synthase	− citrate synthase	− citrate synthase	
			− peroxisomal isocitrate lyase	
			− succinate dehydrogenase	
			− fumarase	
Heterotrophy	50 442 608 modes	49 571 879 modes	4 936 794 modes	157 799 modes
Increase	+ in silico biomass transporter	+ in silico biomass transporter	+ in silico biomass transporter	+ in silico biomass transporter
	+ in silico biomass synthesis	+ in silico biomass synthesis	+ in silico biomass synthesis	+ in silico biomass synthesis
	+ plastidic pyruvate dehydrogenase	+ plastidic pyruvate dehydrogenase	+ plastidic pyruvate dehydrogenase	+ plastidic pyruvate dehydrogenase
Decrease	− CO_2 diffusion	− CO_2 diffusion	− CO_2 diffusion	− CO_2 diffusion
				− ATP synthesis

5.4.2. *A. thaliana* as biotechnological production platform

The emerging significance of plants in biotechnology, now drove our curiosity towards assessing the intrinsic potential of *A. thaliana* as production platform for biotechnologically relevant traits. For this purpose, the biosynthetic pathways of ten interesting chemicals were added to the *Arabidopsis* model (see chapter 4.1) and, for comparison, also to the *Escherichia coli* network from Leighty and Antoniewicz (2012). Elementary flux modes were calculated for the photoautotrophic plant model, whereas *E. coli* was evaluated on glucose as substrate. The number of calculated flux modes, theoretical maximum yield and carbon efficiency for each product were extracted (Fig. 5-11 and Appendix 10.8). For thale cress the feasible region consisted of approximately 5 to 7 million flux distributions for all traits, except for methionine and threonine, displaying a much larger solution space of 12 and 13 million modes, respectively. In general, *E. coli* displayed a much smaller solution space, with less modes and lower yields, as exemplified for cysteine production (Fig. 5-11). As the size of the solution space is directly correlated to metabolic complexity, plants display a much higher intricacy. This implies that changes to single modes or reactions will have less impact on the *in vivo* production, making them more robust to stressors than bacterial production strains. Furthermore, the maximal theoretical product yield in *A. thaliana* was at least as high as the production potential of *E. coli* (Fig. 5-11). Especially for cysteine and methionine a large increase in achievable product yield is witnessed. All traits, except for IPPP production, exhibited a carbon efficiency of more than 80 %. Generally, *Arabidopsis* exhibited a high natural capacity to over-produce each of the studied molecules under autotrophic growth, contributing to the growing industrial appeal of plants as biotechnological factories.

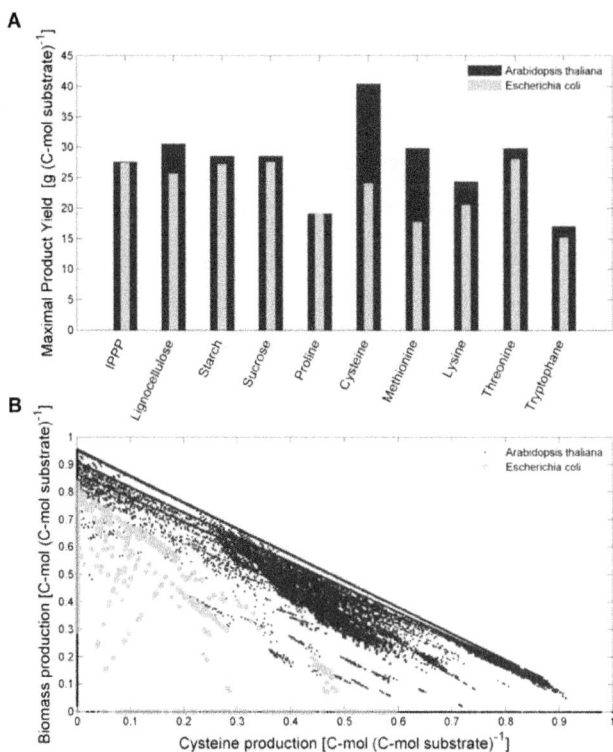

Figure 5-11: Biotechnological potential of *A. thaliana*

(A) Maximal theoretical product yield of ten valuable compounds in *Arabidopsis thaliana* and *Escherichia coli* and (B) visualization of the solution space for cysteine production C-mol (C-mol substrate)$^{-1}$ for both organisms.

Additionally, genetic targets were predicted for ten biotechnologically interesting products. Six different subsets of product forming modes were evaluated separately to search for potential targets that might allow even improved formation (Appendix 10.8), because the inspection of the entire set of modes again did not provide any meaningful results. In general, only few targets could be identified and a consistent increase in predictive power was discovered when further constraining the solution space (Fig. 5-12). For larger subsets, sufficient correlation could only be found for reactions directly linked to either biomass or product formation, whereas the smaller subsets uncovered targets, located mostly in the plastid, but also some in the mitochondrion.

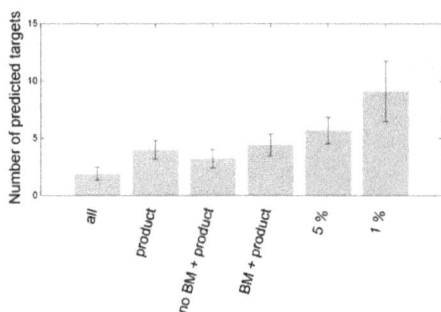

Figure 5-12: Increase in predictive power by focusing on mode subsets

Comparison of prediction potential of different elementary mode subsets used for target identification by Flux Design.

Overall, the simulations revealed that the plant metabolic network of *Arabidopsis* exhibits a high potential for overproduction of added value traits (Fig. 5-11). In contrast to bacteria and fungi (Becker et al., 2011; Driouch et al., 2012), however, the analysis did extract only a small set of potential targets for the plant network. Together with the enormous solution space, this points to a robust nature of plant metabolism. The complex, multi-compartment network provides many alternative flux distributions, which might cause a lower identifiability of genetic targets.

6.   *In vivo* flux investigation of photoautotrophy in rice seedlings

In this chapter a methodology to perform non-stationary ^{13}C-metabolic flux analysis in whole rice plants is described in detail. Construction of a tailor-made flux incubator allowed precise control of temperature, humidity and CO_2 concentration during labeling of whole plants with $^{13}CO_2$. Extensive analytical processing of time-dependent shoot samples delivered large amounts of peak data, which were then automatically converted into their respective mass isotopomer distribution vectors with a specially designed software-tool. Subsequently, the gathered data were fitted to a high-quality genome-based metabolic network for *Oryza sativa*, as established in chapter 4.2. Herewith, the metabolism of the crop *O. sativa* was investigated, allowing elucidation of the *in vivo* intracellular carbon partitioning and of the plants' necessity for futile cycling of resources. Additionally, the effect of imazapyr, an acetohydroxyacid synthase (AHAS) inhibiting herbicide, on rice metabolism was inspected using the newly established workflow. This case-study acts as proof-of-principle, as the known inhibition of branched-chain amino acid synthesis could be confirmed. Moreover, elucidation of intracellular fluxes enabled an additional, deeper understanding of the immediate metabolic effects of the treatment with respect to the alterations in carbon partitioning around the pyruvate node. As this method could also be adopted to other crops and stress inducers, such as abiotic stresses, herbicides, fungicides and many more, it has great potential in green biotechnology.

6.1.   Development of specialized flux incubator for $^{13}CO_2$ labeling experiments

The *in vivo* labeling of plant material for ^{13}C-based metabolic flux analysis should occur under highly controlled, constant conditions, as abiotic perturbations can influence metabolite levels and, thus, fluxes. Therefore, a custom-made flux incubator with integrated temperature regulating and air-humidification system was developed (See chapter 3.6 for technical specifications). The 620 L gas-tight reactor included a ventilating system ensuring sufficient air-circulation for adequate mixing of reactor atmosphere and was connected to an online monitoring system for CO_2 (quadrupole mass spectrometry system). The electrical inflator connected to a specialized CO_2-absorber allowed for the removal of up to 96% of the atmospheric CO_2. Additionally, this set-up allowed temperature, humidity and CO_2 levels to be recorded and maintained at constant levels throughout an entire labeling experiment (Fig. 6-1). It becomes obvious that sufficient mixing of reactor atmosphere exists, as immediately after injection with tracer gas, CO_2 levels were already equilibrated. Additionally, the reactor was equipped with a turning table for better positioning of the plants in front of the sampling port. This allowed a more frequent sampling as plants could be exactly positioned and re-positioned for sampling. The sampling port and tailor-made scissors permitted the fast harvest of individual plants throughout the experiment with minimal perturbation of the reactor environment (Fig. 6-1).

Labeling enclosures for $^{13}CO_2$ application to whole plants have previously been developed that allowed humidity, temperature and CO_2 control (Andersen et al., 1961; Nouchi et al., 1995), such as the commercially available Biobox. However, these set-ups do not allow for sampling of individual plants without perturbing the reactor atmosphere and have therefore exclusively been used for pulse-chase (Römisch-Margl et al., 2007) or label-dilution experiments (Huege et al., 2007). The latter can be applied for isotopic non-stationary labeling experiments, however requires extensive periods of expensive tracer application. Both previously reported labeling experiments for subsequent isotopic nonstationary flux estimation, applied individual experimental runs for every time point and replicate in labeling chambers without control of the set conditions (Ma et al., 2014; Szecowka et al., 2013). In contrast, our tailor-made flux incubator allowed the joint incubation of all plants belonging to one experiment under extensive control of abiotic conditions, which is of great value to the quality and reproducibility of the generated data.

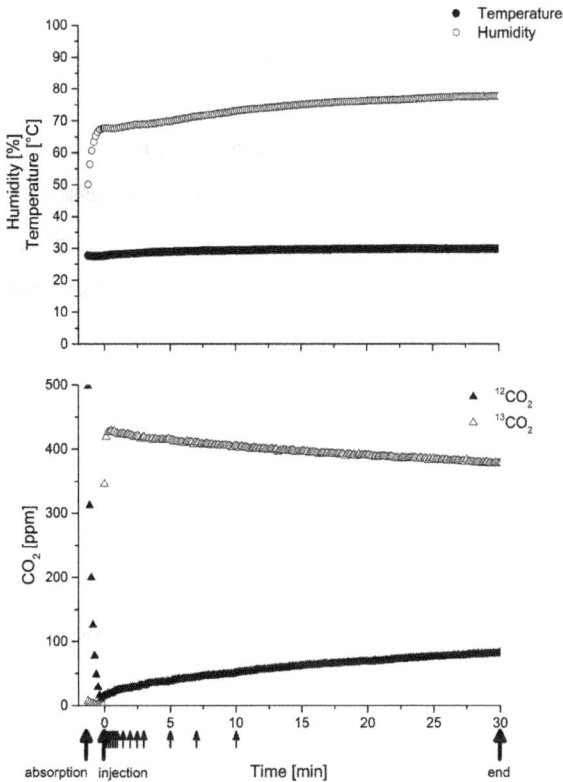

Figure 6-1: Atmospheric conditions in the flux reactor during a $^{13}CO_2$ labeling experiment

Temperature (set to 28 °C) and relative humidity (set to 75 %) were monitored during a $^{13}CO_2$ labeling experiment. Additionally, both $^{12}CO_2$ and $^{13}CO_2$ were measured using an integrated online MS. Prior to the start of the experiment, absorption takes place for 60 s, which diminishes $^{12}CO_2$ levels to below 25 ppm. Immediately after, 400 ppm of $^{13}CO_2$ are injected into the flux reactor, which indicated the start of the experiment. The large arrows indicate the beginning and end of both the absorption and the experiment, whereas the smaller arrows represent the 15 sampling time points.

6.2.   Experimental setup and data-acquisition

In vivo metabolic flux analysis through ^{13}C-tracing requires extensive data coverage, including information on the applied tracer, biomass composition, uptake- and production rates, as well as transient isotopomer distributions. Optionally, pool sizes can be obtained to further enrich the data set. A wide variety of state-of-the-art analytical tools were applied to compile the required data (Table 6-1), counting several mass spectrometry platforms such as GC-MS, LC-MS/MS, GC-irMS, EA-irMS and online quadrupole MS. This allowed the quantification and/or qualitative monitoring of several metabolite classes, ranging from direct biomass constituents such as proteinogenic amino acids and lipid esters, to intracellular metabolites such as sugar-phosphates and organic acids.

Tracer information. To accurately infer fluxes from ^{13}C incorporation patterns in plant metabolism, it is essential to examine the mixture of substrate precisely, as all labeling originates from the applied ^{13}C-tracer substrate. Therefore, both $^{13}CO_2$ and $^{12}CO_2$ were monitored during each labeling experiment by a highly sensitive online quadrupole MS (Fig. 6-1). After absorption, the substrate mixture in the flux reactor consisted of less than 25 ppm $^{12}CO_2$ and approximately 440 ppm $^{13}CO_2$. During the course of the 30 minute lasting experiment, $^{12}CO_2$ levels rose steadily to roughly 75 ppm at the end of the experiment and the $^{13}CO_2$ concentration dropped to 390 ppm. This observation most likely originates from $^{13}CO_2$ assimilation and respiration of previously assimilated $^{12}CO_2$. Assuming that plant enzymes do not differentiate between isotopologues, both $^{12}CO_2$ and $^{13}CO_2$ are respired and re-assimilated, which significantly impacts the reactor atmosphere. In turn, the continuously changing mixture of substrates effects the transient isotopomer distributions and was therefore included in the modeling.

Table 6-1: Overview on data acquisition for ^{13}C-INST-MFA.

Data set	Applied methods	Analytical coverage
Tracer information	online MS	$^{12}CO_2$ and $^{13}CO_2$
Biomass composition	C/N combustion	protein content
	quantitative LC-MS/MS	amino acids and some organic acids
	quantitative GC-MS	organic acids
	quantitative photometry	starch and free sugars
	FAME analysis	fatty acids
	literature search	nucleic acids, pigments and cell wall composition
Uptake and production rates		
• Biomass yield	EA-irMS	$^{13}CO_2$
	Gravimetrical analysis	shoot biomass
• Export to root	GC-irMS	amino acids and sugars
	quantitative GC-MS	amino acids
	quantitative photometry	sugars
Isotopomer distributions	GC-MS analysis	sugars, sugar-phosphates, organic acids and amino acids
	LC-MS/MS analysis	
	GC-irMS analysis	
Pool sizes	quantitative GC-MS	sugars, sugar-phosphates, organic acids and amino acids
	quantitative LC-MS	

Biomass composition. In exponentially growing tissues, most of the assimilated carbon is directed towards anabolism. Because the macromolecular composition of biomass determines the anabolic demand for specific biomass precursors (see chapter 4.2 for more detailed information), the major fractions of shoot biomass were determined analytically to achieve an overall biomass composition for rice seedlings (Fig. 4-5 and Appendix 10.1). By enriching the measured fractions with selected primary literature, 85.5 % of carbon could be recovered. The remaining 14.5 % represented the seedlings ash-content, with literature values between 13 and 15 % (Misra et al., 2006). This rather exact match can be taken as a measure of quality. From the macromolecular composition and the molar anabolic precursor demand for each biomass component (Appendix 10.2 and 10.4), the total carbon requirement for synthesis of one gram of biomass could be calculated (Table 6-2). Apparently, carbon is mainly channeled to anabolism through glucose 6-phosphate, phospoenolpyruvate and erythrose 4-phosphate towards macromolecular proteins, starch and cell wall components, respectively.

Table 6-2: Anabolic precursor demand for fifteen-day old rice seedlings. All values are expressed as C-mmol $(gDW)^{-1}$. Abbreviations can be taken from Appendix 10.5.

Precursor	3PG	AcCoA	AKG	E4P	F6P	G6P	GAP	OAA	P5P	PEP	PYR	SUCCoA
Anabolic demand	1.333	1.620	3.920	4.242	1.313	7.732	0.324	3.004	2.461	6.363	3.838	0.164

Biomass yield. Next, the carbon flow towards shoot anabolism was quantified by integrating gravimetrical analysis with EA-irMS measurements (Fig. 6-2). Over a period of 11 days, between day 11 and 22, plants were harvested and their dry weight (DW) determined gravimetrically. An excellent linear correlation was found between the natural logarithm of these values and time, indicating that rice seedlings grow perfectly exponential with a growth rate of 0.011 h^{-1}. However, the extracted value considered both day-time and night-time metabolism, which have been shown to differ dramatically (see Chapter 5). Therefore, a day-time growth rate of 0.026 h^{-1} was found to more accurately represent biomass formation at the time of the experiment. The latter growth rate was calculated from gravimetrical samples belonging to one light-phase. Additionally, the CO_2 uptake rate had to be determined to derive the biomass yield. For this purpose, EA-irMS analysis was performed for 15 plants during a 30 minute labeling experiment. A linear increase in $^{13}CO_2$ concentration in the plant tissue, enabled quantification of the CO_2-uptake rate, which was 0.00803 mmol CO_2 (g DW)$^{-1}$. By dividing the growth rate with the CO_2-uptake rate, a biomass yield of 6.59 g DW (100 mmol CO_2)$^{-1}$ was calculated, which amounts to a biomass carbon yield of 2.68 g C in DW (100 C-mmol CO_2)$^{-1}$, when considering a C (w/w)% of 40.8 % as determined by C/N combustion analysis.

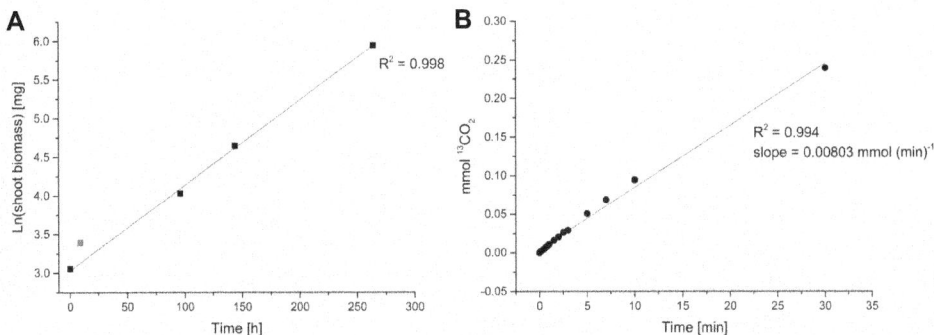

Figure 6-2: Gravimetrical analysis of shoot biomass and quantification of CO_2 uptake rate

(A) Weight (mg) of *O. sativa* seedling shoots between day 11 and 21 after sowing. Linear regression delivered an overall specific growth rate of 0.011 h^{-1}. The first two data points were additionally consulted to extract a day-time specific growth rate of 0.026 h^{-1}. (B) Amount of $^{13}CO_2$ incorporated in seedling shoots during a 30 minute labeling experiment. Linear regression revealed a CO_2 uptake rate of 0.00803 mmol (min)$^{-1}$.

Export to the root. In addition to carbon assimilation into shoot biomass, several photosynthetic assimilates are exported to support root metabolism. The major form of carbon exported to the root is sucrose (Chapter 4.2). Therefore, the measured root sucrose concentration and the MID of sucrose throughout the labeling experiment were included in the model. This enabled deduction of the sucrose export flux.

Transient isotopomer distributions. By including LC-MS/MS, GC-MS and GC-irMS analytical

platforms, MID vectors of a large variety of internal metabolites could be determined. LC-MS/MS

focused mainly on sugar-phosphates (F6P, G6P, FBP, P5P, RBP and S7P), sugar acids (2PG,

3PG, 6PG, GLYCO) and other organic acids (PEP, PYR), whereas GC-MS was able to

determine the isotopomer distribution of amino acids (alanine, aspartate, glutamate, isoleucine,

leucine, phenylalanine, proline, valine, serine (2 fragments) and threonine), organic acids (FUM,

AKG, CIT, MAL, PYR, GLYCER (2 fragments), SUCC), G6P and sucrose. In addition, GC-irMS

delivered enrichment data on several amino acids (alanine, aspartate, asparagine, glutamate,

glutamine, glycine, lysine, serine, tyrosine and valine), sugars (sucrose (shoot and root),

glucose, fructose (2 fractions) and inositol) and malate. GC-irMS enrichment values are far less

informative for flux determination than isotopomer distributions, however, in this way, labeling

tracking in metabolites such as glycine and glucose could at least partially be fulfilled. Overall,

many different metabolites were captured that were distributed evenly across the metabolic

network. For more information on the analytics and the MIDs of the measured metabolites, the

reader is kindly referred to the partner-dissertation of this work (Dersch, 2016).

Independent of the analytical platform, label incorporation was accurately captured with very

small error bars and high reproducibility among replicates (Fig. 6-3). Strong label incorporation

was observed for metabolites from the CBB cycle, PP pathway and EMP pathway, whereas only

minor ^{13}C incorporation was displayed by most amino acids, soluble sugars and metabolites of

the TCA cycle. Noteworthy enrichment was additionally found for sucrose, alanine, serine,

glycine, aspartate and phenylalanine. A combination of large metabolite turnover and small pool

size could explain the observed fast enrichment, on the other hand, small metabolite turnover

and large pool size could lead to barely detectable label incorporation. Consequently, sampling

time-points needed to be chosen carefully, to capture the isotopic transient profile of as many

metabolic intermediates as possible. For this purpose, many samples were taken early during

the labeling experiment, with increasing time-intervals towards the later phase.

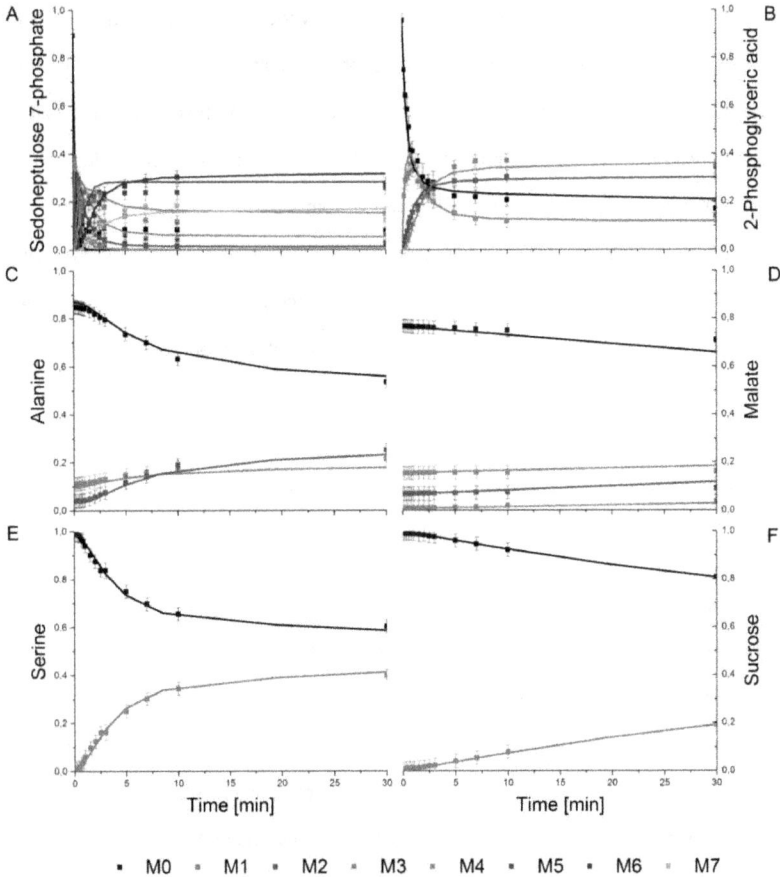

· M0 ▫ M1 · M2 ▪ M3 ▫ M4 · M5 · M6 ▫ M7

Figure 6-3: Isotopomer distribution and enrichment of selected metabolites from central carbon metabolism determined by LC-MS/MS, GC-MS and GC-irMS analysis

(A) Transient MID of sedoheptulose 7-posphate, measured by LC-MS/MS (B) Transient MID of 2-phosphoglyceric acid, measured by LC-MS/MS (C) Transient MID of alanine, measured by GC-MS (D) Transient MID of malate fragment, measured by GC-MS (E) Enrichment of serine, measured by GC-irMS (F) Enrichment of sucrose, measured by GC-irMS. All measurement data were overlain with the simulated values for the best-fitting flux distribution.

In total, 35 MIDs and 18 enrichments, accounting for more than 200 raw MS spectra, were measured per sample. For one experiment with five replicates and 15 time-points, a totality of nearly 15 000 spectra needed to be analyzed and converted to their MIDs and enrichments, respectively. The availability of automated integration protocols facilitated the spectral analysis, however automated conversion of integrated peak areas to their respective MIDs and enrichments was needful. For this purpose a MATLAB-based graphical user interface (GUI) was developed that automatically converted peak areas from LC-MS/MS, GC-MS, GC-irMS, online MS and EA-irMS platforms to the required output format (Fig. 6-4). The presented GUI was highly flexible as it permitted the number of replicates, time points, measured metabolites and platform to be adjusted to specific needs. Additionally, it allowed for optional automated correction of natural labeling, which is of great value during quality control of the measurements. The provided output formats included immediate visualization in MATLAB, export to MS Office Excel and export to an INCA-compatible format. Processing of 15 000 peaks could be performed within minutes, which is a tremendous improvement over manual labor.

Pool sizes. The metabolite concentration, also known as its pool size, provides valuable information for isotopically instationary flux determination, as the actual amount of incorporated label, and thus the flux, can be derived directly from the metabolites labeling profile and its pool size. However, accurate measurement of small intracellular molecules is error prone due to rapid conversion and leakage (Young et al., 2011). As labeling information on its own provides sufficient information for the flux estimating algorithm to converge to the optimal flux distribution, pool sizes are most often omitted from the model. In this case, pool sizes are variables that are estimated in the process. As the proposed metabolic network for rice seedlings is rather large and complex, it should be considered if inclusion of direct pool size measurements potentially enhances flux identifiability (See Chapter 6.4). Therefore, pool sizes were measured for many amino acids, sugars and some intermediates from central carbon metabolism, using GC-MS and LC-MS/MS techniques (Appendix 10.10). Most measured metabolic pools were very small

(< 2 µmol (g DW)$^{-1}$), however, conspicuous amounts of glutamate/glutamine

(115 ± 7 µmol (g DW)$^{-1}$), aspartate/asparagine (18 ± 0.5 µmol (g DW)$^{-1}$), serine

(7 ± 0.2 µmol (g DW)$^{-1}$), alanine (10.7 ± 0.8 µmol (g DW)$^{-1}$), glycine (3.6 ± 0.3 µmol (g DW)$^{-1}$),

glucose (13 ± 2 µmol (g DW)$^{-1}$), fructose (19 ± 1 µmol (g DW)$^{-1}$) and 3PG (42 ± 3 µmol (g DW)$^{-1}$)

were detected. As 3-phosphoglycerate is the entry-point for newly assimilated carbon, it is not

surprising that this metabolite occurs plentifully. Furthermore, glutamate and aspartate are

known long-distance transport amino acids (Dersch et al., 2016b), whereas serine, alanine and

glycine have an additional metabolic role in photorespiration, next to their anabolic function,

which explains their high concentrations.

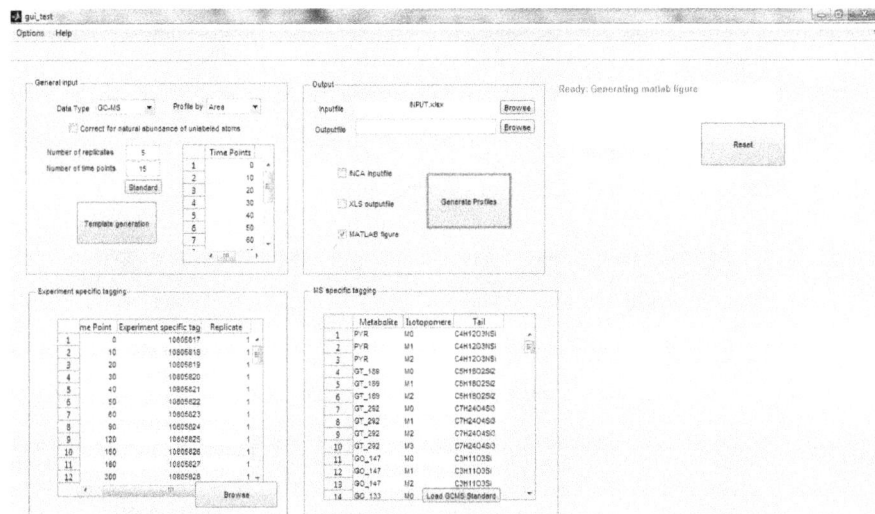

Figure 6-4: Screen shot of GUI for the automated conversion of peak data into MIDs

GC-MS data of an experiment with 5 replicates and 15 time points was processed, without correcting for natural isotope abundance. The experiment-specific-tagging window delivers information on the measurement identities, whereas in the MS-specific-tagging window information about the measured isotopomers is required. The latter can be entered by hand or standard information can be uploaded. In addition, this GC-MS standard information can be updated to your needs for future computations. After selecting the desired output format, the MID profiles can be generated. Both for the individual replicates as for their average, an MID profile, including error bars, will be rendered.

A comprehensive dataset for ^{13}C-INST-MFA was now available that included online MS measurements for accurate description of the applied tracer, extensive analytical data on biomass composition, the biomass yield derived from gravimetrical and EA-irMS analyses, sucrose labeling profile and concentration for the root to determine sucrose export, as well as transient MID information on 53 internal metabolites from multiple MS platforms. In addition, pool size information on 27 metabolic pools was available to attempt further flux identifiability.

6.3. Iterative optimization of network topology

The above established data-set allowed iterative optimization of network topology to assure maximum flux identifiability. In this regard, different network topologies were considered, before deciding on the final model characteristics. Firstly, the necessity of a parallel EMP and oxidative PP pathway was contemplated, after which compartmentation was reconsidered. Further refinement acknowledged the level of detail in photorespiration and sugar metabolism. The resulting optimized network topology was described in detail in Chapter 4.2.

6.3.1. Compartmentation is essential to accurately represent *in vivo* metabolism

A first impression on fluxes was delivered for the fully compartmentalized network topology depicted in Fig. 6-5. The very low residual error, which is far outside the statistically accepted range (Table 6-3), is a first indication that the parallel model is too complex to be resolved with the available dataset. Furthermore, nearly all fluxes had large errors, of which 45 standard deviations exceeded 50 % of the flux value (Appendix 10.9.3). Both the EMP pathway and the oxidative PP pathway operate in parallel in the cell, more specifically in the cytosol and in the plastid (Plaxton, 1996), with transporters in place for many metabolites (Facchinelli & Weber, 2011; Sweetlove & Fernie, 2013). Consequently, many alternative routes are available to the plant and as a result, more effort is required to unambiguously identify the *in vivo* metabolic fluxes. As no compartment-specific labeling data was available for the above-mentioned pathways, the introduction of an artificial compartment, comprising both the plastid and the cytosol, constrained the solution space significantly. Unfortunately, this completely unparalleled network model (Fig. 6-6) also did not deliver a statistically accepted flux distribution (Table 6-3).

Table 6-3: SSR for the different tested network topologies and their 95% confidence intervals.

Network topology	SSR $\left[\chi^2_{\alpha/2}(n-p), \chi^2_{1-\alpha/2}(n-p)\right]$
parallel (original parallel network)	1625.0 [1972.0 2225.8]
unparallel (completely unparalleled network, final network without F6P.cp)	3203.2 [1960.3 2213.4]
noHexokinase (final network without hexokinase function)	2590.7 [2005.0 2260.8]
WT (final network)	1956.6 [2004.0 2259.8]

In this case, the residual error was much larger than expected, indicating that the completely unparalleled model was an oversimplified version of reality, thus some form of compartmentation is required to accurately represent *in vivo* metabolism. Some of the metabolites of the EMP pathway also participated in other pathways that were strictly confined to one specific compartment. Examples include 3-phosphoglycerate in photorespiration and fructose 6-phosphate in the PP pathway, both occurring only in the plastid. As a bi-directional 3-phosphoglycerate (3PG) transporter between plastid and cytosol was previously discovered, the labeling of plastidic 3PG should not differ from the labeling found in the cytosol.

Figure 6-5 (next page): Original network topology with parallel EMP and oxidative PP pathways

The here presented network topology has strictly separated plastidic and cytosolic EMP and oxidative PP pathways, with many plastidial membrane transporters in place. This is the topology with the highest complexity, and therefore lowest identifiability, as many alternative metabolic routes are available.

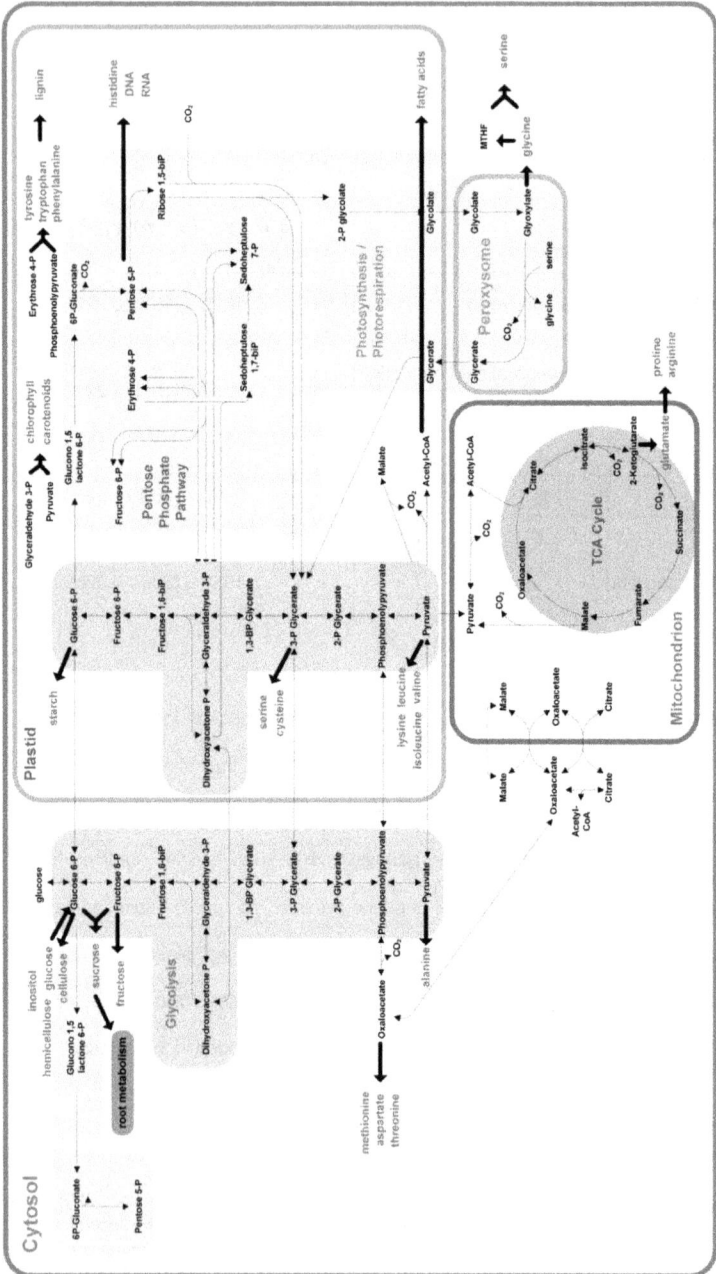

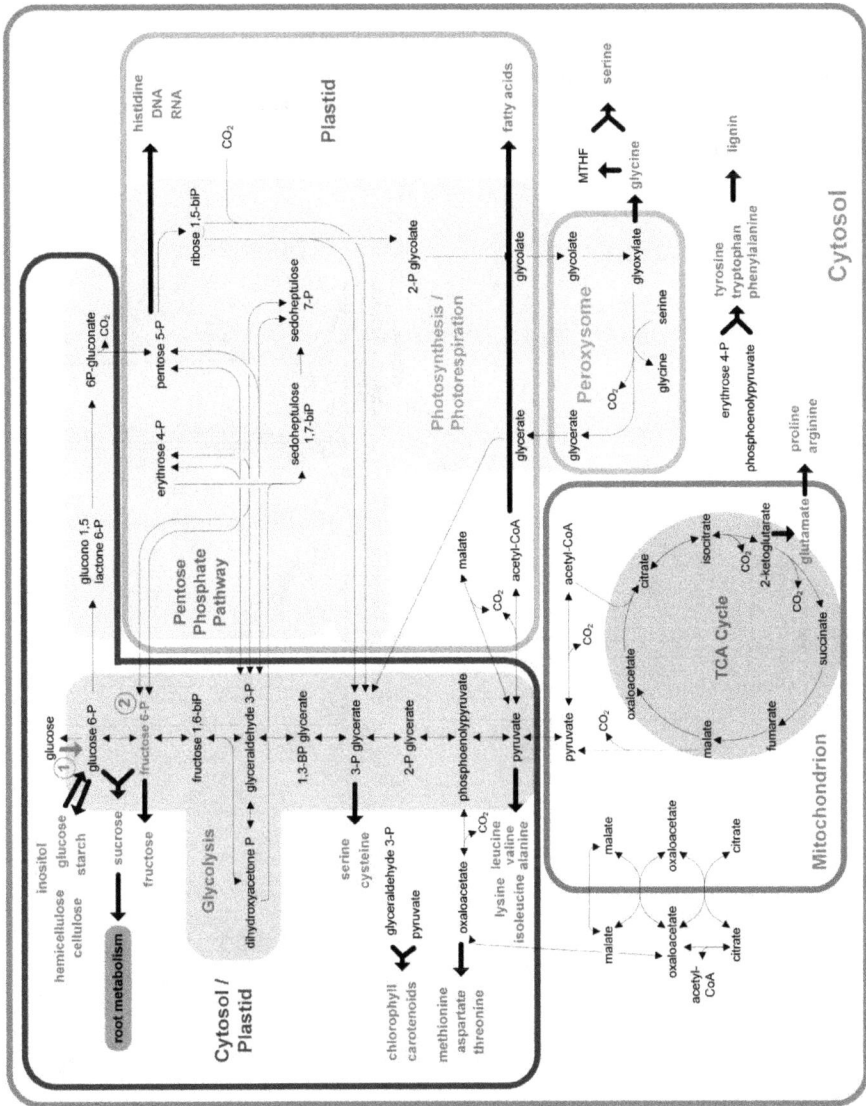

Figure 6-6 (previous page): Alternative network topologies with strictly unparalleled EMP and oxidative PP pathways or omission of the hexokinase function

The here presented network topologies vary from the one shown in Fig. 4-4 by either (1) omitting the hexokinase function or (2) by uncompartmentalizing fructose 6-phosphate.

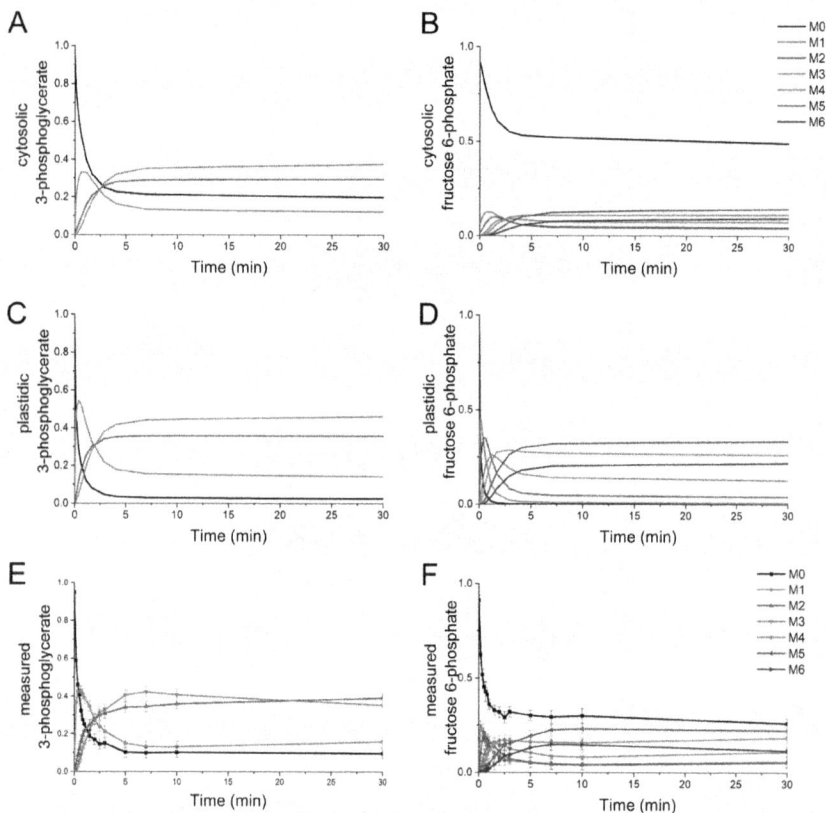

Figure 6-7: Simulated and measured MIDs for 3-phosphoglycerate and fructose 6-phosphate

Simulated compartment specific transient mass isotopomer distributions of (A) cytosolic and (B) plastidic 3-phosphoglycerate and (D) cytosolic and (E) plastidic fructose 6-phosphate. (C,F) Measured isotopomer distributions for both metabolites (not compartment specific) from LC-MS/MS analysis. Figure originates from joint effort with Lisa Dersch, Institute of systems biotechnology, Saarland University.

This hypothesis was reinforced by the near-to perfect fit of 3PG modeled isotope profiles to the measured data (Fig. 6-7). However, so far, only unspecific transporters for fructose 6-phosphate (F6P) between cytosol and plastid have been proposed (Facchinelli & Weber, 2011; Plaxton, 1996; Taiz & Zeiger, 2006). If a bi-directional transporter would be present, again stationary plastidic F6P labeling should not differ from the F6P labeling in the cytosolic compartment. However, large discrepancies were found in their labeling patterns (Fig. 6-7), indicating that two spatially strictly separated pools of F6P, without direct exchange, were present in the cell. This theory was supported by the measured F6P profile, which represented a mixture of cytosolic and plastidic F6P. Thus, compartmentation of specific pathways is essential to describe the observed labeling dynamics. Therefore, the final network topology considered two spatially separated F6P pools, one in the artificial cytosolic-plastidic compartment and one in the plastid (Fig. 4-4). This reduced the residual error to within acceptable limits of the 95 % confidence interval (Table 6-3). In addition, the absolute errors on the calculated fluxes were reduced significantly during this optimization process (Fig. 6-8).

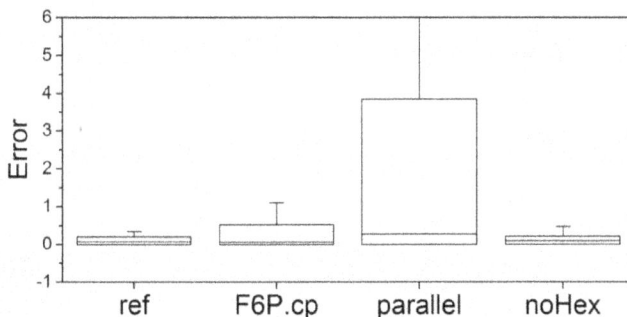

Figure 6-8: Absolute errors on net fluxes for the different network topologies

Box plot representations of the absolute error on the net fluxes for the selected network topologies. ref (final reference network), F6P.cp (final network without compartmentation of F6P or completely unparalleled network), noHex (final network without hexokinase), parallel (original parallel network). The data shown were taken from Appendix 10.9.3.

6.3.2. Sugar/sugar-phosphate substrate cycling also occurs in autotrophic tissues

A similar optimization procedure as the one described above considered sugar metabolism. In the original model, glucose 6-phosphate (G6P) was the precursor for irreversible glucose synthesis and thus, G6P could only be produced by phosphoglucose isomerase from F6P. However, clear differences in stationary labeling levels (Fig. 6-9) show that G6P has an additional precursor-pool that clearly has a lower stationary labeling level than the F6P pool. The model without the hexokinase function was incapable of rendering a statistically accepted flux distribution (Table 6-3), however, due to the reduced solution space, absolute flux errors were lower (Fig. 6-8). The unequivocal cause for the high residual error was an ill-fitted glucose 6-phosphate MID (Fig. 6-10). Addition of a hexokinase as additional way of synthesizing glucose 6-phosphate, improved the fit between measured and simulated G6P and glucose transient labeling profiles considerably (Fig. 6-10), asserting the hypothesis that futile substrate cycling takes place between glucose and G6P.

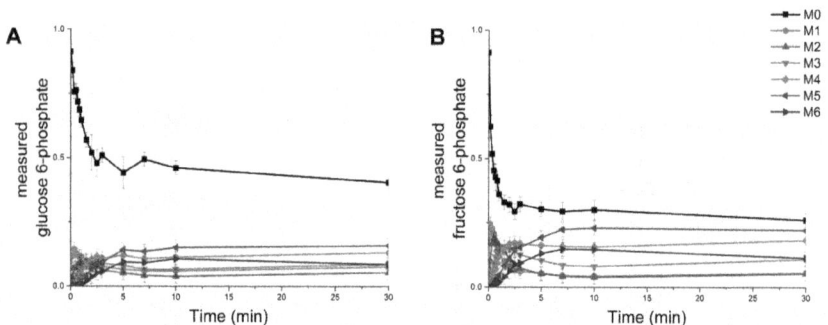

Figure 6-9: Measured MID profiles for (A) glucose 6-phosphate and (B) fructose 6-phosphate from LC-MS/MS

Similar to the proposed glucose-to-glucose 6-phosphate turnover, a high glucose 6-phosphate-to-glucose turnover was discovered in *Zea mays* root tips from *in vivo* labeling experiments (Alonso et al., 2005). The authors showed that in addition to glucose recycling from storage compounds (vacuole, starch, sucrose and cell wall polysaccharides), a recycling of glucose from previously synthesized glucose 6-phosphate took place. Thus, futile cycling at the cost of one ATP per cycle occurred. Although this study involved heterotrophic root tips, it is not farfetched that a similar process might occur in the opposite direction in autotrophic tissues. The necessary (unlabeled) glucose for such futile cycling could originate from large amounts of stored glucose in the plant vacuoles (Poschet et al., 2011; Szecowka et al., 2013; Tohge et al., 2011) and/or from glucose recycling from storage compounds such as starch (Baroja-Fernández et al., 2001). The function of such futile cycling is yet unknown, however one may speculate that it is involved in Pi remobilization (Plaxton & Tran, 2011; Raghothama, 1999) or regulation of glycolysis through ATP utilization (Urbanczyk-Wochniak et al., 2003).

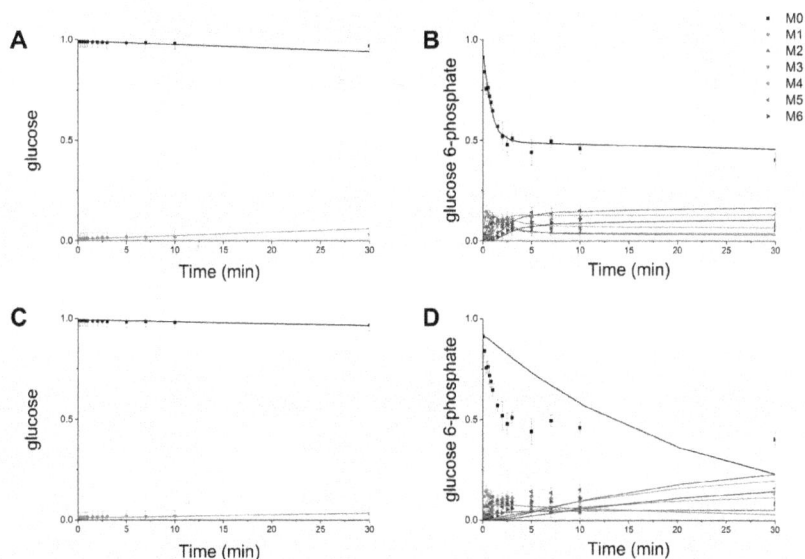

Figure 6-10: Simulated and experimental MIDs for glucose and G6P

Simulated transient mass isotopomer distributions for glucose (A+C) and glucose 6-phosphate (B+D), overlaid with their respective measurements (A-B) when the model included a hexokinase and (C-D) when the hexokinase was excluded.

The final network model was organized into four compartments, with the EMP and oxidative PP pathways located in the cytosolic-plastidic compartment. However, compartmentation of the F6P pool was included, based on the observed labeling profiles. In addition, to account for the observed glucose-to-glucose 6-phosphate cycling a hexokinase was included.

6.4. High flux-resolution achieved by comprehensive data coverage

The previously established dataset, which included transient isotopomer information on 53 internal metabolites, 27 pool sizes, biomass composition and yield, sucrose export to the root and the available tracer composition, was iteratively fitted against the simulated values from the optimized network topology to infer a best-fitting flux distribution. The calculated fluxes had narrow confidence intervals (Appendix 10.9.4) and fitted the measured data excellently (Fig. 6-3 and Table 6-4). Next, it was investigated if the use of such comprehensive dataset really improved flux resolution dramatically or whether certain subsets of data could be omitted with minimal loss in identifiability.

6.4.1. Improved flux identifiability through refined photorespiratory pathway

The labeling pattern in amino acids is of high value for flux elucidation in stationary metabolic flux analysis (Wittmann, 2007), therefore 22 isotopically non-stationary labeling profiles of 15 different amino acids were included in the model. Comparison of the SSR (Table 6-4) and absolute errors between the model with and without amino acid information (Fig. 6-11), clearly demonstrates that including amino acid measurements dramatically improved flux identifiability. Especially the photorespiratory pathway, CBB cycle, PP pathway and EMP pathway benefitted from amino acid information. This was possibly due to the more detailed representation of photorespiration. Metabolic flux analysis is usually focused on the investigation of central carbon metabolic conversions in the cell and lumps anabolic pathways. However, photorespiration is strongly entwined with the anabolic pathways for glycine and serine biosynthesis. Frequently, this linkage with anabolism is neglected and the reactions are lumped (Ma et al., 2014; Young et al., 2011), however incorporation of amino acid labeling profiles required the incorporation of more detail in photorespiration to accurately represent metabolism. In turn, this improved overall flux identifiablity.

6.4.2. Measured pool sizes as lower boundaries decrease flux errors

It was previously shown that direct pool size measurements are superfluous to achieve high flux identifiability (Young et al., 2011). This is a major advantage as quantification of intracellular pools tends to be error prone due to leakage. Consequently, measured pool sizes are often underestimated, which in turn leads to ill-fitting flux scenarios. However, excluding direct pool size data from the parameter optimization increases the number of variables that can be varied, leading to more uncertainty on the calculated fluxes. Nevertheless, instead of adding the pool size data as direct parameter measurements, they could be implemented as fixed lower boundaries, thus potentially reducing the parameter uncertainty. In this way, net fluxes in the proposed highly complex network could be identified with less uncertainty (Fig. 6-11). Especially the photorespiratory pathway, the PP pathway, the CBB cycle and the EMP pathway showed significant improvement of absolute flux error, which is probably linked to the high propensity of pool size measurements in these pathways.

Table 6-4: SSR for the different subsets of data and their 95% confidence intervals.

Data coverage	SSR $\left[\chi^2_{\alpha/2}(n-p), \chi^2_{1-\alpha/2}(n-p)\right]$
WT (final network with full data coverage)	1956.6 [2004.0 2259.8]
noAA (final network without MS data on amino acid metabolism)	1669.2 [1419.6 1636.2]
noGCMS (final network without GC-MS data)	1173.3 [1010.9 1194.9]
noPool (final network without pool size measurements)	1950.0 [2004.0 2259.8]

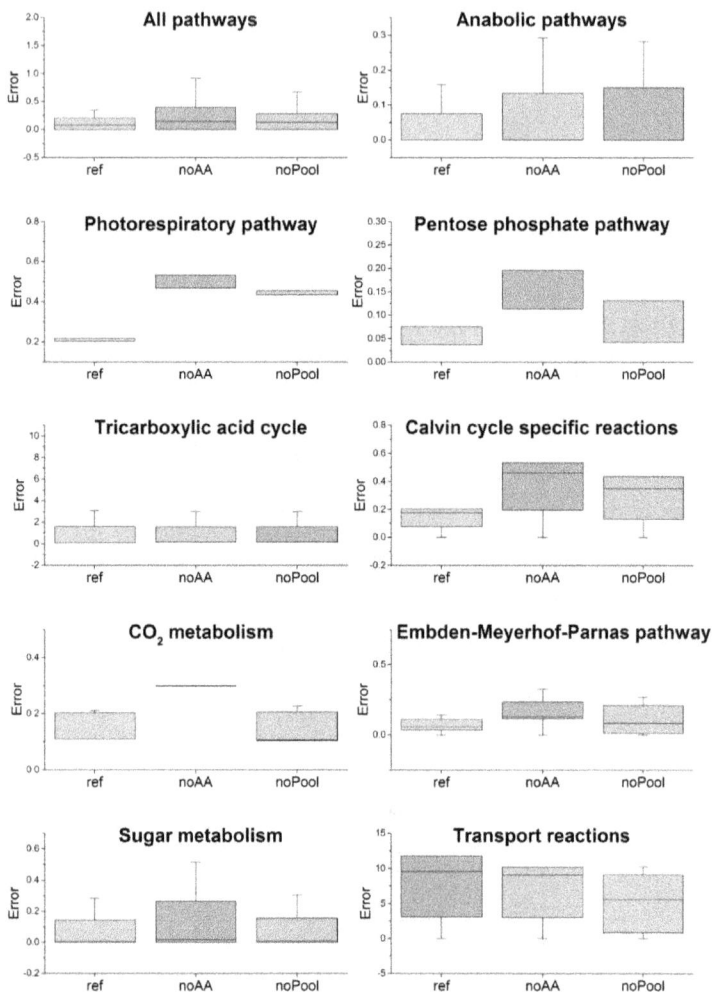

Figure 6-11: Absolute errors on net fluxes calculate with different data sets

Box plot representations of the absolute error on the net fluxes for the selected data set. ref (final reference network with full data coverage), noAA (final network without MS data on amino acid metabolism), noGCMS (final network without GC-MS data), noPool (final network without pool size measurements). Green box plots signify the best, orange the second best and red the worst data set of the three. The errors shown were taken from Appendix 10.9.2.

6.5. Photoautotrophic metabolism of rice seedlings reveals energy dissipation and redox shuttling through futile carbon cycling

Firstly, an experimental workflow was established that allowed simultaneous labeling of all seedlings belonging to one experiment and thus, increased reproducibility (Chapter 6.1). Secondly, the analytical platform was developed that allowed measurement of no less than 53 transient isotopomer profiles (Appendix of Dersch (2016)), 27 intracellular concentrations (Appendix 10.10), biomass composition (Appendix 10.1 and Table 6-2) and yield (Fig. 6-2), sucrose export to the root (Fig. 4-6) and the available tracer composition (Fig. 6-1). Thirdly, this extensive dataset was used to optimize network topology to achieve highest flux resolution (Chapter 6.3). Now, five replicate experiments, each with 15 rice seedlings (15 days old), were performed in which plants were labeled with $^{13}CO_2$. Analytical processing delivered the comprehensive data set that was then used to infer a best-fitting flux distribution from iteratively fitting simulated values to the experimentally determined data set. The rendered flux values were considered to be the *in vivo* metabolic fluxes in rice seedlings (Fig. 6-12). Repetition of flux optimization from multiple alternative starting points, ensured that a global minimum was reached. In addition, statistical analysis by parameter continuation was performed on the 95 % confidence level, which revealed narrow confidence intervals across the entire data set (Appendix 10.9.1), with only few exceptions around the pyruvate node. Together with the excellent agreement between simulated and measured data (Fig. 6-3 and Table 6-4), this now enabled the exploration of whole rice plants *in vivo* with high confidence (Fig. 6-12).

Figure 6-12 (next page): Fluxes in a wild type rice seedling from ^{13}C-INST-MFA

Metabolic flux distribution for wild type *Oryza sativa* L. *ssp. japonica* seedlings under ambient conditions as calculated by ^{13}C-NMFA. The value on the arrow and the thickness of the arrow represent the flux value. All fluxes are normalized to 100 mmol h^{-1} $^{13}CO_2$ uptake, therefore fluxes can be seen as percentage of substrate uptake. Standard errors are derived from the estimated 95 % confidence interval as determined by parameter continuation. The arrows into metabolites displayed in green visualize the amount of building blocks needed for anabolic synthesis.

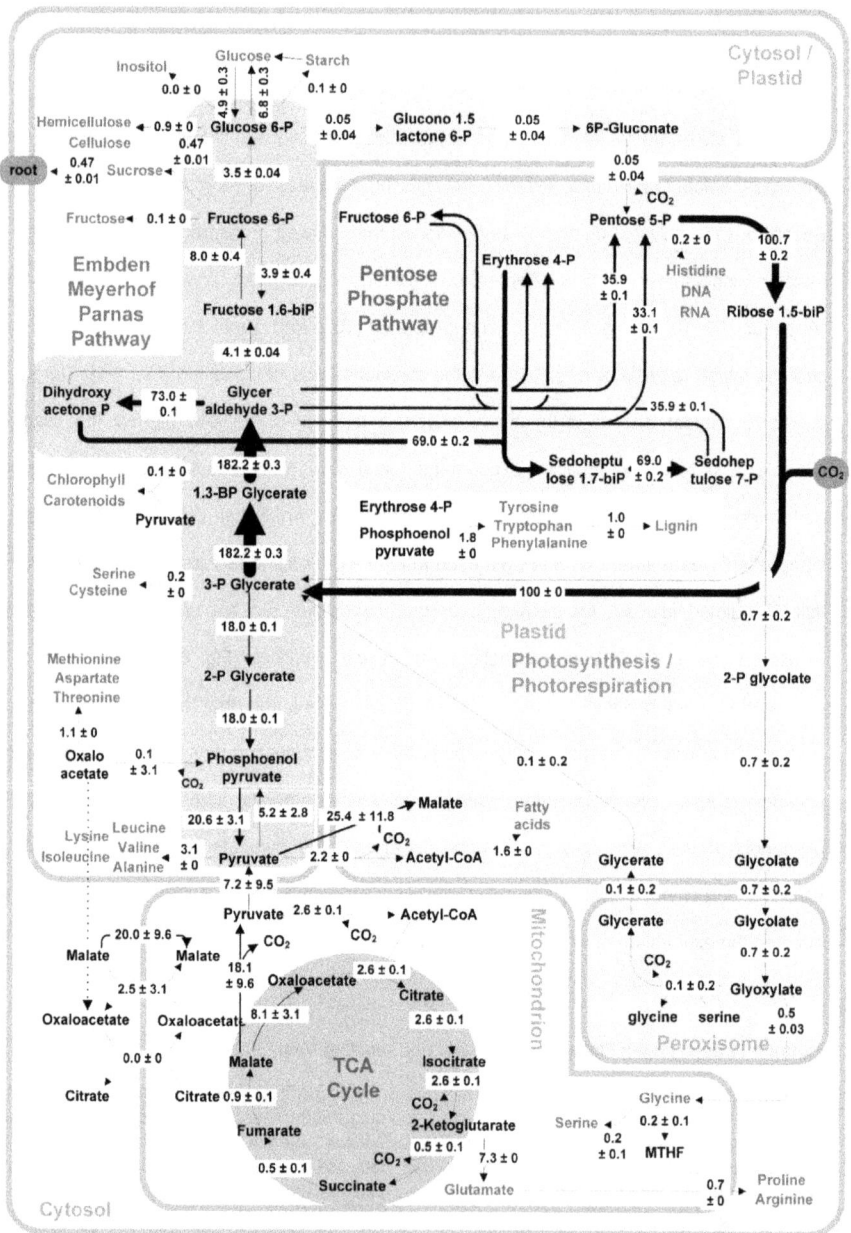

Cytosol / Plastid

Inositol
0.0 ± 0

Glucose ◄--- Starch
4.9 ± 0.3
6.8 ± 0.3

Glucose 0.1 ± 0

Hemicellulose ◄ 0.9 ± 0 Glucose 6-P
Cellulose 0.47
0.47 ± 0.01
root 0.47 ± 0.01 Sucrose◄ 3.5 ± 0.04

0.05 ± 0.04 ► Glucono 1.5 lactone 6-P 0.05 ± 0.04 ► 6P-Gluconate

0.05 ± 0.04

Fructose◄ 0.1 ± 0 Fructose 6-P

Fructose 6-P ◄

Pentose 5-P 0.2 ± 0 100.7 ± 0.2

CO₂

Embden Meyerhof Parnas Pathway

8.0 ± 0.4
3.9 ± 0.4
Fructose 1.6-biP
4.1 ± 0.04

Erythrose 4-P

Histidine 35.9 ± 0.1
DNA
RNA 33.1 ± 0.1 Ribose 1.5-biP

Pentose Phosphate Pathway

Dihydroxy acetone P ◄ 73.0 ± 0.1 Glycer aldehyde 3-P

35.9 ± 0.1

69.0 ± 0.2

Chlorophyll 0.1 ± 0 182.2 ± 0.3
Carotenoids ◄ 1.3-BP Glycerate
Pyruvate
182.2 ± 0.3

Sedoheptu lose 1.7-biP 69.0 ± 0.2 ► Sedohep tulose 7-P

Erythrose 4-P
Phosphoenol 1.8 pyruvate ± 0 Tyrosine Tryptophan Phenylalanine 1.0 ± 0 ► Lignin

CO₂

Serine 0.2 Cysteine ± 0 3-P Glycerate ◄

100 ± 0

0.7 ± 0.2

18.0 ± 0.1

Methionine
Aspartate 2-P Glycerate
Threonine
▲
1.1 ± 0

Plastid
Photosynthesis / Photorespiration

2-P glycolate 0.7 ± 0.2

18.0 ± 0.1

Oxalo 0.1 ► Phosphoenol
acetate ± 3.1 CO₂ pyruvate
20.6 ± 3.1 5.2 ± 2.8 25.4 ± 11.8 ► Malate
Lysine Leucine CO₂
Isoleucine Valine 3.1 Pyruvate 2.2 ± 0 ► Acetyl-CoA
Alanine ± 0 7.2 ± 9.5

0.1 ± 0.2 0.7 ± 0.2

Fatty acids 1.6 ± 0

Glycerate Glycolate 0.1 ± 0.2 0.7 ± 0.2

Pyruvate 2.6 ± 0.1 ► Acetyl-CoA

20.0 ± 9.6
Malate CO₂ CO₂
Malate
18.1 2.6 ± 0.1
2.5 ± 3.1 ± 9.6 Oxaloacetate Citrate
Oxaloacetate Oxaloacetate 8.1 ± 3.1 2.6 ± 0.1
0.0 ± 0 Malate **TCA** Isocitrate
Citrate Citrate 0.9 ± 0.1 **Cycle** 2.6 ± 0.1
Fumarate CO₂
0.5 ± 0.1 2-Ketoglutarate
CO₂ 0.5 ± 0.1 7.3 ± 0
Succinate ◄ Glutamate

Mitochondrion

Glycerate Glycolate 0.1 ± 0.2 0.7 ± 0.2
CO₂ 0.7 ± 0.2
0.1 ± 0.2 Glyoxylate
glycine serine 0.5 ± 0.03

Peroxisome

Glycine ◄ 0.2 ± 0.1
Serine ◄ 0.2 ± 0.1 MTHF
0.7 ± 0 ► Proline Arginine

Cytosol

High activity of plastidic metabolism. The CBB cycle and non-oxidative PP pathway appear highly active with fluxes of 182.2 ± 0.3 mol% for glyceraldehyde 3-phosphate dehydrogenase, 69.0 ± 0.2 mol% for sedoheptulose 1,7-bisphosphatase and 35.9 ± 0.1 mol% for transketolase. Here, all fluxes are normalized to 100 mmol of CO_2 uptake. Thus, a flux value of 182.2 mmol $(100 \text{ mmol } CO_2)^{-1}$ can be interpreted as 182.2 mol% of the substrate uptake flux. The CBB cycle and non-oxidative PP pathway are strongly entwined and are responsible for the direct fixation of carbon from CO_2 by the catalytic action of RuBisCO (100 mol%). Carbon assimilation is a pivotal process for plant survival, which explains the observed high activities.

Gluconeogenic and glycolytic function of EMP pathway. As 3PG is the first metabolite that contains assimilated CO_2, it is interesting to learn that it is being dispersed both in a gluconeogenic and a glycolytic direction by the EMP pathway. The EMP pathway operated in a gluconeogenic direction between 3PG and G6P, towards starch, sucrose and cell wall synthesis, with fluxes of 3.5 ± 0.04 mol% for phosphoglucose isomerase and 182.2 ± 0.3 mol% for glyceraldehydes 3-phosphate dehydrogenase. A glycolytic way occurred from 3PG to pyruvate, $(18.0 \pm 0.1$ mol%) to fuel mitochondrial metabolism and amino acid and fatty acid biosynthesis. The observed partitioning of carbon at the 3PG node seems crucial to provide the plant with the necessary anabolic building blocks.

Small, but significant fluxes through the oxidative pentose phosphate pathway and photorespiratory pathway. Fluxes through photorespiration (0.7 ± 0.2 mol%) and the oxidative PP pathway (0.05 ± 0.04 mol%) appear near-to inexistent. However, label incorporation in glucose 6-phosphate, pentose 5-phosphate and 6-phosphogluconate reveals that although the enrichment in 6-phosphogluconate is significantly lower than in the surrounding metabolites, still a distinct labeling pattern can be observed. This indicates that the fluxes through the oxidative pentose phosphate pathway are small, but existent (Fig. 6-13). Similarly, label incorporation in glycolate is significantly lower than in its precursor metabolite ribulose 1,5-bisphosphate.

Nonetheless, a distinct labeling pattern can be observed, which again leads to the conclusion that the fluxes through the photorespiratory pathway are small, but existent (Fig. 6-14). Interestingly, photorespiration is involved in the reduction of wasteful products of RuBisCO oxygenation, but potentially also plays a role in pathogen defense (Sørhagen et al., 2013) and abiotic stress response (Voss et al., 2013). Therefore, it is admissible to state that the investigated wild type rice seedlings are not subjected to any form of abiotic or biotic stress.

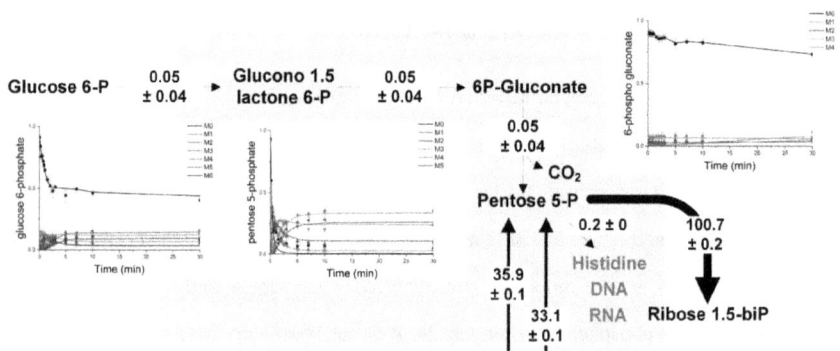

Figure 6-13: Distinct label incorporation in the oxidative pentose phosphate pathway

Label incorporation in glucose 6-phosphate, pentose 5-phosphate and 6-phosphogluconate reveals that although the enrichment in 6-phosphogluconate is significantly lower than in the surrounding metabolites, still a distinct labeling pattern can be seen. This indicates that the fluxes through the oxidative pentose phosphate pathway are small, but existent. Fluxes were taken from Fig. 6-12.

Most carbon directed towards anabolism. About 94 % of the assimilated carbon was directed towards shoot growth, whereas the sucrose export of 0.47 ± 0.01 mol%, or 6 C-mol%, fueled root metabolism. Previously, carbon translocation to the root was quantified through enrichment of root material in a pulse-chase experiment (Dersch et al., 2016a). This straight-forward method reported a carbon efflux of 6 C-mol%, which matches the value from our, more detailed, analysis perfectly. This congruency puts emphasis on the quality of our simulations and gives confidence to the attained conclusions. Nearly all assimilated carbon was fixed in either shoot or root biomass formation. As a result, CO_2 loss through respiration was very low (0.19 ± 0.11 mol%).

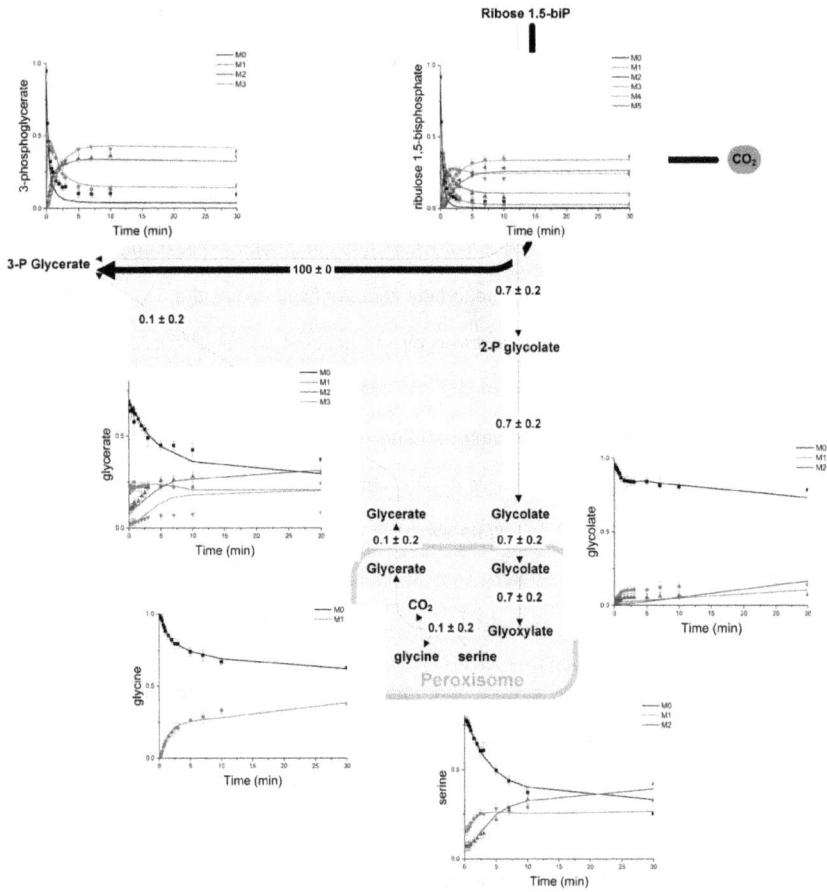

Figure 6-14: Distinct label incorporation in the metabolites of photorespiration

Label incorporation in glycolate is significantly lower than in its precursor metabolite ribulose 1,5-bisphosphate. Nonetheless, a distinct labeling pattern can be seen. This indicates that the fluxes through the photorespiratory pathway are small, but existent. Fluxes were taken from Fig. 6-12.

Anabolic fluxes drive particular pathways. As most of the carbon was used for anabolism, it is now interesting to have a closer look at the largest anabolic sinks. Biosynthesis of plant material required mostly glucose 6-phosphate (20.7 C-mol%) and phosphoenolpyruvate (17.1 C-mol%) for the synthesis of cell wall components (cellulose and hemicellulose) and particular amino acids, respectively. Furthermore, protein synthesis originated largely from erythrose 4-phosphate (11.4 C-mol%), pyruvate (10.3 C-mol%), α-ketoglutarate (10.5 C-mol%) and oxaloacetate (8.1 C-mol%), whereas fatty acid synthesis required 4.3 C-mol% acetyl-CoA. Fascinatingly, many pathways were driven by the anabolic demand of its intermediates. For instance, the synthesis of different sugars and cell wall components were responsible for the gluconeogenic activity of the EMP pathway. Furthermore, the TCA cycle mainly displayed an anabolic function towards the synthesis of glutamate, arginine and proline, with fluxes of 2.6 $\pm$ 0.1 mol% and 0.5 $\pm$ 0.1 mol% for citrate synthase and succinate dehydrogenase, respectively. The anabolic control of flux is intriguing, as in heterotrophic organisms fluxes are mainly controlled through fuelling, i.e. the provision of redox power and energy.

Exchange of redox power across the internal organelle membranes. In contrast to stoichiometric approaches, ^{13}C-based metabolic flux analysis did not rely on underlying assumptions concerning redox balancing. Therefore, in addition to providing valuable insights into *in vivo* carbon metabolism, it was also possible to investigate redox metabolism. An interesting question at this point is, how reducing equivalents are transported across the internal membranes. The main process to generate NADPH (i.e. photosynthesis) occurs in the plastid, whereas the TCA cycle in the mitochondrion provides the cell with NADH. These processes are specialized functions that are restricted to particular organelles, whereas the cellular demand for NADH and NADPH is not. As NAD(P)H is incapable of directly crossing the cellular membranes, it must be transported using shuttles, such as the DHAP/3PG translocator in the plastidic membrane or the MAL/OAA shuttle, located both in the plastidic and mitochondrial membrane (Hoefnagel et al., 1998). Unfortunately, OAA is highly unstable and therefore it is very difficult the accurately measure its MID profile. In addition, malate is a highly connected node, participating in many different metabolic processes, with little ^{13}C incorporation. This renders it difficult at this stage to infer accurate flux estimations for the MAL/OAA shuttle. However, it is interesting to note that an alternative way of redox channeling could be identified *in vivo*. Imported malate was decarboxylated to pyruvate in the mitochondrion, thus producing mitochondrial NADH. After export of pyruvate to the cytosol/plastid, reversible NADP-dependent malic enzyme consumed NADPH, while producing malate that could, in turn, be transferred into the mitochondrion. In this way, redox power was channeled from the plastid into the mitochondrion through futile carbon cycling, again indicating the TCA cycle has a mere anabolic function. So far, the existence of a PYR/MAL shuttle in the mitochondrial membrane of plants has not been recognized (Grafahrend-Belau et al., 2008; Haferkamp et al., 2011; Klingenberg, 2008; Laloi, 1999; Picault et al., 2004), however pancreatic cell lines have shown a cycling of pyruvate and malate across the mitochondrial membrane, which is very similar to that proposed here (Jensen et al., 2008; MacDonald, 1995). This MAL/PYR shuttle likely operates in addition to the MAL/OAA shuttle. From this it becomes clear that cellular transporters play a central role in compartmented metabolism.

Substrate cycling lowers ATP:NAD(P)H ratio. The MIDs of glucose and glucose 6-phosphate already indicated that significant substrate cycling occurred (see above). Such substrate cycling is energetically unfavorable as is comes at a net ATP cost of 1 mol per substrate cycle. As flux confidence was particularly high in sugar metabolism, the proposed futile cycle could be quantified with high confidence (4.5 ± 0.3 mol%). The function of such futile cycling is yet unknown, however one may speculate that it is involved in regulation of glycolysis through ATP utilization (Urbanczyk-Wochniak et al., 2003) or in meeting different metabolic demands for ATP. It emerged from *in silico* plant modeling that already small variations in the ratio of photosynthetic ATP:NADPH synthesis around the accepted *in vivo* value of 1.5, cause dramatic changes in specific fluxes (Chapter 5). Moreover, these changes were mainly involved in dissipating excess energy or redox power. It is now observed that such dissipation processes also occur *in vivo* as the glucose-to-glucose 6-phosphate cycling effectively lowers the ATP:NAD(P)H ratio of metabolic energy demand from 1.67 to 1.65. Also other ATP-dissipating futile cycles could be identified, such as the PYR-to-PEP cycling (5.2 mol%), lowering the ratio further to 1.62. Plant metabolism appears capable of adapting to changing ATP:NAD(P)H ratios. In this context, one might now contemplate if it is the plasticity in photosynthetic electron flow that matches cellular requirements (Allen, 2003), or if plant metabolism is adjusted to deal with imposed ATP:NAD(P)H ratios.

Strong similarity to optimal fluxes from EFM analysis. To provide additional understanding of the observed PYR/MAL shuttle and ATP-dissipating substrate cycling, *in silico* elementary flux mode analysis was performed (Fig. 6-15). For this purpose, the metabolic network model was extended with redox and energy metabolites, as well as with the photosynthetic light reactions (Appendix 10.12). The general distribution of flux was very similar between the *in vivo* fluxes from ^{13}C-INST-MFA and those from *in silico* EFM analysis (Fig. 6-16). A similarly high activity of the plastidic metabolism, carbon partitioning at the 3PG node, anabolic function of the TCA cycle and low photorespiration were observed. Overall, only few fluxes deviated significantly (Fig. 3-16): (1) CO_2 loss because the *in vivo* growth rate is slightly smaller than the theoretical maximum (2) futile substrate cycling through plastidic malic enzyme, malate export to the mitochondrion, mitochondrial malate dehydrogenase and pyruvate transport to the plastid (3) futile substrate cycling between phosphoenolpyruvate and pyruvate through activity of pyruvate kinase and pyruvate, diphophate kinase (4) futile substrate cycling between fructose 6-phosphate and fructose 1,6-bisphosphate through activity of fructose 1,6-bisphosphatase and 6-phosphofructokinase (5) futile substrate cycling between glucose and glucose 6-phosphate through activity of hexokinase and glucose 6-phosphatase. From this we learn that the previously identified substrate cycles are decomposable and not elementary, thus are truly futile. The lack of futile cycling in the stoichiometric fluxes emphasizes the importance of ^{13}C-based methods, as it is clear that regulatory levels exist that can not be elucidated with stoichiometric

approaches alone.

Figure 6-15 (next page – left): Fluxes in a wild type rice seedling from EFM analysis

Metabolic flux distribution for wild type *Oryza sativa* L. *ssp. japonica* seedlings as calculated by EFM analysis. The value on the arrow and the thickness of the arrow represent the flux value. All fluxes are normalized to 100 mmol h^{-1} CO_2 uptake, therefore fluxes can be seen as percentage of substrate uptake. The arrows into metabolites displayed in green visualize the amount of building blocks needed for anabolic synthesis.

Figure 6-16 (next page – right): Fluxes in a wild type rice seedling, determined by ^{13}C-INST-MFA and EFM analysis

The metabolic phenotype of rice seedlings, determined by ^{13}C-INST-MFA is compared to the respective elementary flux mode yielding maximum biomass. Fluxes are given in mmol $(100 \text{ mmol } CO_2)^{-1}$. For visualization purposes the identity line is given (full line) with its 90 % confidence interval (dashed line). Only very few differences were observed: (1) CO_2 loss (2) futile substrate cycling through plastidic malic enzyme, malate export to the mitochondrion, mitochondrial malate dehydrogenase and pyruvate transport to the plastid (3) futile substrate cycling between phosphoenolpyruvate and pyruvate through activity of pyruvate kinase and pyruvate, diphophate kinase (4) futile substrate cycling between fructose 6-phosphate and fructose 1,6-bisphosphate through activity of fructose 1,6-bisphosphatase and 6-phosphofructokinase (5) futile substrate cycling between glucose and glucose 6-phosphate through activity of hexokinase and glucose 6-phosphatase.

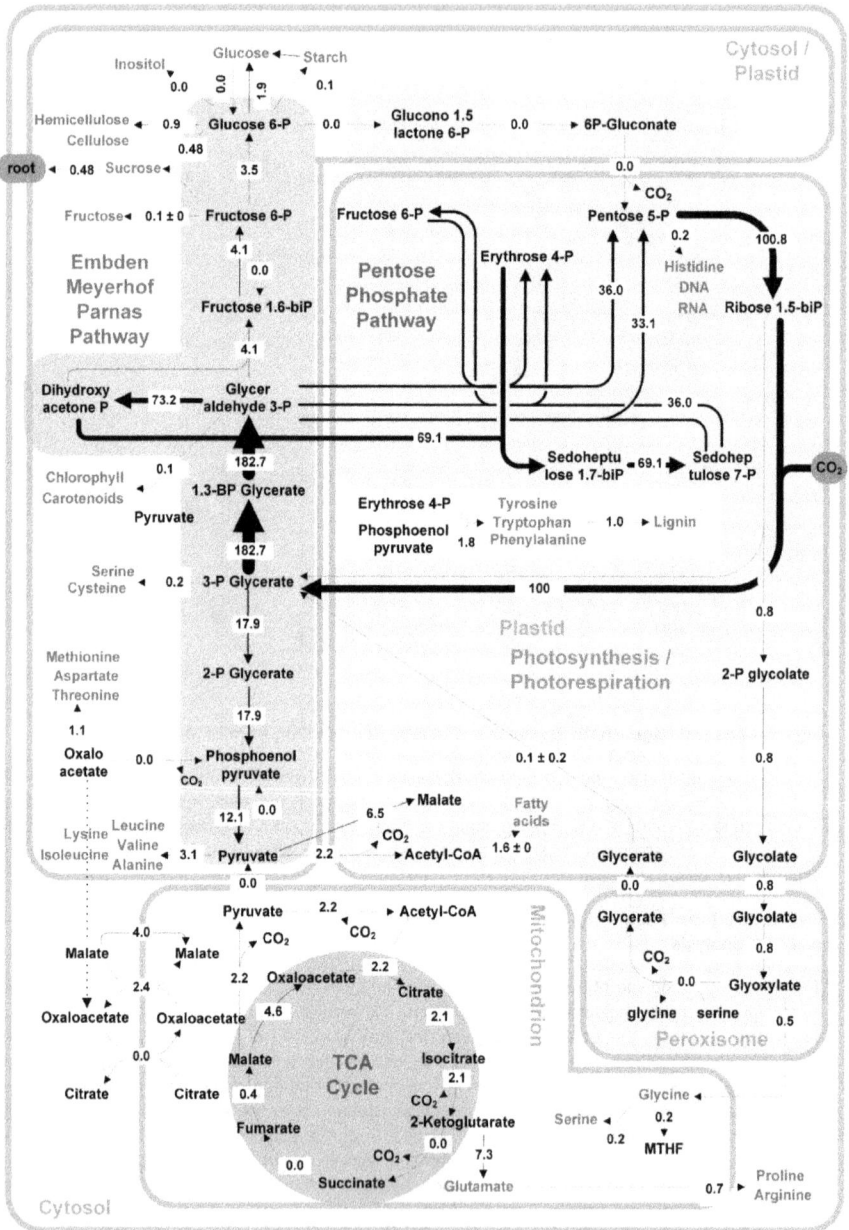

A

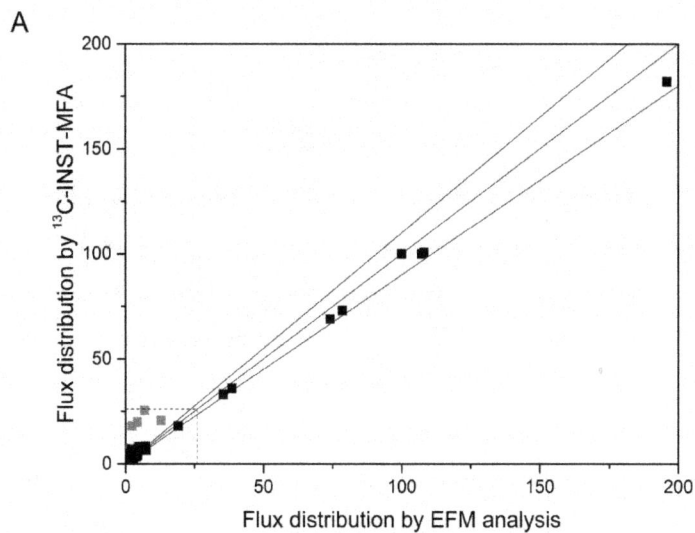

B

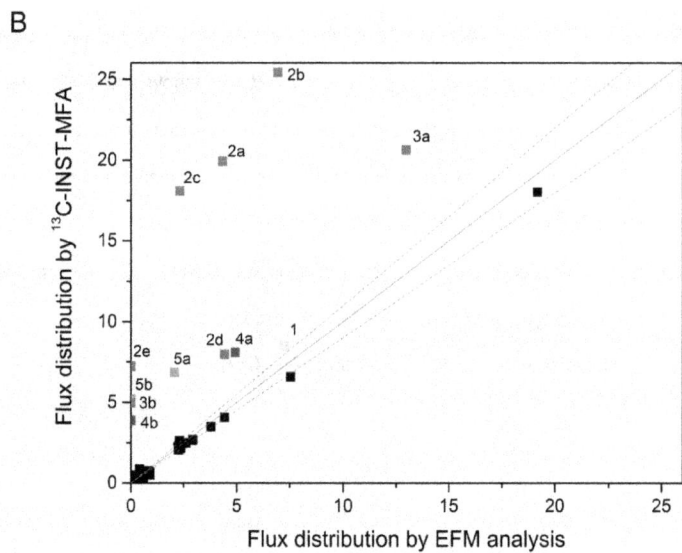

6.6.   Deepening insights into imidazolinone-herbicide effect on rice

Now that a workflow has been established that enabled detailed analysis of *in vivo* metabolic pathways in whole plants, a first case-study could be performed to evaluate its future application potential. For this purpose, ^{13}C-INST-MFA was performed under treatment with Imazapyr, a well known herbicide affecting the branched chain amino acid biosynthesis (Appendix 10.9.2). As Imazapyr is a slow-acting herbicide (Shaner & Singh, 1991), the experiment was performed only 4 h after treatment to ensure uncompromised viability and to increase the likelihood of capturing the immediate metabolic response. To exclude possible metabolic effects of the solvent, DMSO-treated seedlings were chosen as a reference. In total 150 rice seedlings (15 days old) were used in 10 labeling experiments, i.e. five for each treatment. For each of the conditions, comprehensive data sets were compiled with 53 transient isotopomer profiles (see Appendix of Dersch (2016)), 27 intracellular concentrations (Appendix 10.10), biomass composition (Appendix 10.1), sucrose export to the root and the available tracer composition (See Appendix of Dersch (2016)). Best-fitting flux distributions, derived from iterative fitting of the experimental data set, represented the *in vivo* fluxes in rice seedlings under imazapyr and DMSO treatment, respectively (Appendix 10.9.2). Multiple starting points for the optimization ensured that a global minimum was reached, in which an excellent agreement between simulated and measured data was achieved (Imazapyr: SSR = 1842.7 [1824.7 – 2069.1], DMSO: SSR = 2013.6 [2004.0 – 2259.8]). Parameter continuation revealed narrow confidence intervals, which were similar to those obtained in the wild type (Appendix 10.9.2). When subjecting the observed fluxes to a student's t-test, significant changes could be identified (Fig. 6-17).

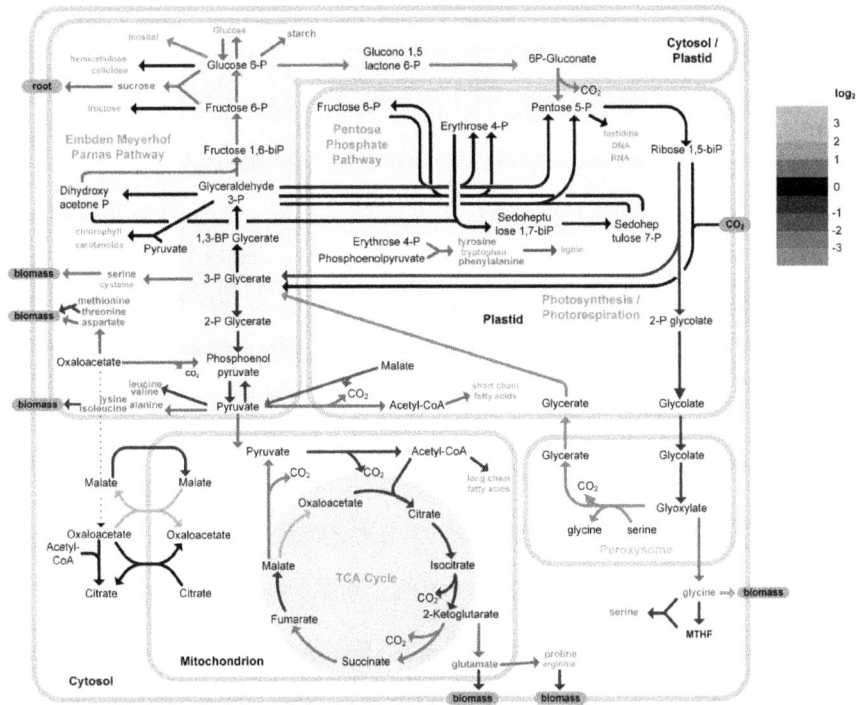

Figure 6-17: Metabolic flux changes imposed by herbicide-treatment

Differences in metabolic flux distribution between imazapyr-treated *Oryza sativa* seedlings and DMSO-treated seedlings as a reference. Green and red arrows indicate a significant increase and decrease in flux under imazapyr treatment, respectively. Also, changes in intracellular pool sizes of amino acids, sugars and starch are indicated by the same color code. Differences in fluxes were considered significant based on a standard student's t-test ($\alpha=0.05$).

Localized effects of Imazapyr on sugar metabolism and the pyruvate node. The most apparent effects of Imazapyr on primary metabolism were located in the mitochondrion, around the PEP and PYR nodes (Fig. 6-18) and in sugar metabolism (Fig. 6-19). Mitochondrial malate dehydrogenase showed a 4.5-fold increase, whereas succinate dehydrogenase and oxoglutarate dehydrogenase exhibited decreases of 4.9 and 4.8-fold, respectively. The observed increase in carbohydrate synthesis was directly linked to augmented precursor supply through the upper EMP pathway of 1.04-fold. In addition, significant changes were observed in the oxidative PP pathway (5.9-fold) and the photorespiratory pathway (1.1-fold), however these fluxes are relative low (< 1.1 mol%) and might not significantly impact stress physiology. The inferred intracellular pool sizes also displayed distinct changes. Sucrose and starch levels were notably increased (1.1-fold and 1.8-fold, respectively), as were all amino acid concentrations, except for those of leucine and valine, which showed strong depletion of 35-fold and 2.6-fold, respectively. Alanine and proline accumulated the most, with an observed fold-change of 1.9 and 3.3, respectively. Imazapyr is an industrially relevant herbicide (Duggleby et al., 2008) that has been thoroughly investigated. Multiple studies have provided evidence that Imazapyr is a potent AHAS inhibitor, a key enzyme in the biosynthesis of branched-chain amino acids (Chang & Duggleby, 1997; Duggleby et al., 2008; Vega et al., 2012). This could also be inferred from the observed reduced flux towards valine and leucine, which demonstrates that *in vivo* ^{13}C-INST-MFA can be successfully applied to elucidate toxic mode of action in whole plants.

Amino acid inhibition influences anabolism through protein turnover and growth inhibition. Next to identifying the mode of action, flux analysis can partake in elucidating the observed macromolecular changes on the molecular level. Although most intracellular amino acids accumulated inside the cell, their respective *de novo*-synthesis was significantly decreased. This points to an increased protein turnover to regenerate the depleted branched-chain amino acids, valine and leucine (Fig. 6-18). The continuous degradation of already existing proteins is likely responsible for the observed increase in all other soluble amino acid concentrations. This is congruent with the macromolecular observations of Royuela et al. (2000) and Shaner and Singh (1991). In turn, this implies an upcoming growth arrest, as is observed after extended application of Imazapyr (Shaner & Singh, 1991). In comparison to the other intracellular amino acids, especially alanine (1.9-fold) and proline (3.3-fold) are strongly increased. The latter is known as an important compatible solute in plant stress response (Liang

et al., 2013), whereas the former could be contributed to a compensatory flux due to pyruvate
overflow (Fig. 6-18).

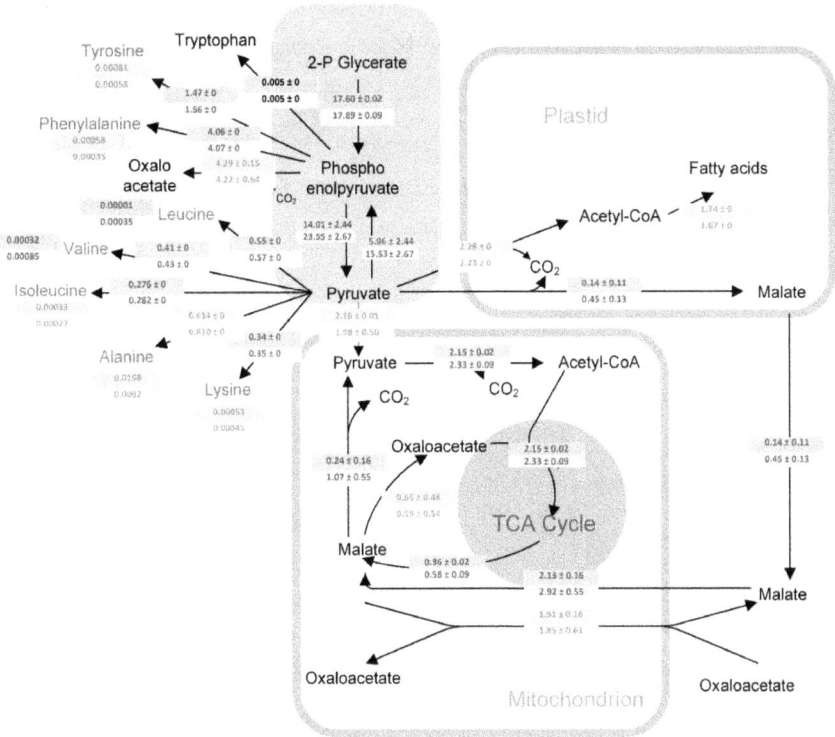

**Figure 6-18: Metabolic flux changes around the pyruvate and phosphoenolpyruvate node imposed
by herbicide-treatment**

Differences in metabolic flux distribution between imazapyr-treated *Oryza sativa* seedlings (upper values)
and DMSO-treated seedlings (lower values) as a reference. Flux values are given on the respective arrow
in mol%. Green and red boxes indicate a significant increase and decrease in flux under imazapyr
treatment, respectively. Also, changes in intracellular pool sizes of amino acids are indicated by the same
color code. Differences in fluxes were considered significant based on a standard student's t-test ($\alpha=0.05$).

Increase in storage pools. Possibly, a compensatory flux, similar to the flux towards alanine, causes the observed increase in short chain fatty acid synthesis from plastidic acetyl-CoA, after decarboxylation of pyruvate by pyruvate dehydrogenase (Fig. 6-18). It is however interesting to note, that only growth-independent macromolecules, such as soluble amino acids, free sugars, starch and free fatty acids, are increased, which again points in the direction of an impending growth arrest. This is corroborated by a previously observed decrease in dry weight, although starch and soluble sucrose levels increased dramatically (Royuela et al., 2000). Obviously, the plant tries to store the assimilated carbon in the most practical way.

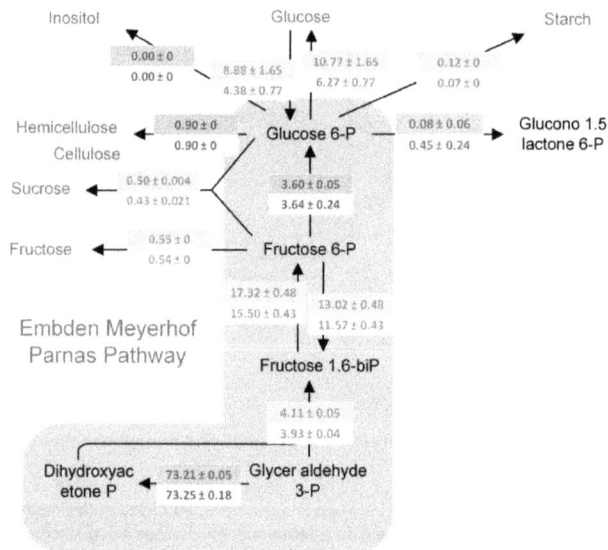

Figure 6-19: Metabolic flux changes in sugar metabolism imposed by herbicide-treatment

Differences in metabolic flux distribution between imazapyr-treated *Oryza sativa* seedlings (upper values) and DMSO-treated seedlings (lower values) as a reference. Flux values are given on the respective arrow in mol%. Green and red boxes indicate a significant increase and decrease in flux under imazapyr treatment, respectively. Differences in fluxes were considered significant based on a standard student's t-test (α=0.05).

Adenylate homeostasis jeopardized due to reduction of growth. It was previously postulated that the observed storage carbohydrate accumulation was caused by reduced phloem loading in combination with decreased export to the root (Chao et al., 1994; Kim & Vanden Born, 1996). However, flux calculations show significantly increased sucrose export from 0.43 ± 0.01 mol% in the reference to 0.50 ± 0.01 mol% under herbicide treatment. This is consistent with previous observations of accumulating carbohydrates in roots (Zabalza et al., 2004). Therefore, it is more likely the increase in sugar metabolism is related to a change in carbon partitioning and ATP requirement, rather than by impaired phloem loading (Chao et al., 1994; Kim & Vanden Born, 1996) or reduced sink strength (Zabalza et al., 2004). The plants appear to store the assimilated carbon preferably as sucrose and starch, when carbon flow towards biomass formation is impaired (Fig. 6-19). This increase in carbohydrate synthesis under Imazapyr treatment and the observed translocation require more ATP than in the control, which is potentially beneficial for the cell. Under normal growth, much cellular ATP is consumed during the *de novo* synthesis of amino acids. As the *de novo* synthesis of many amino acids is reduced (Fig. 6-17), much less ATP will be needed, which could potentially threaten adenylate homeostasis (Noctor 2000). This could be partially counteracted by increased ATP-demand in sugar metabolism. A possibly vulnerable adenylate status is also visible in the increase of ATP dissipation through futile cycling between glucose and glucose 6-phosphate as compared to the control (Fig. 6-19). In addition, the early response of genes encoding ATP-binding cassette transporters, points to the involvement of ATP in the Imazapyr-effect (Manabe et al., 2007).

Reduced demand for redox power. The net import of malate in the mitochondrion is reduced from 1.07 ± 0.55 mol% to 0.22 ± 0.16 mol% under Imazapyr treatment (Fig. 6-18). This decreases the net import of redox power through the pyruvate/malate shuttle by 4.5-fold, which indicates a decreased necessity for NADH in the mitochondrion. This is possibly associated with the decreased anabolic demand for NAD(P)H, due to impaired *de novo* amino acid synthesis. In addition, the reduced anabolism could also drive the observed decrease in TCA cycle activity. In this regard, the diminished cellular demand for redox power, might be related to a possible role for cytochrome P450 in the early transcriptional response to Imazapyr-treatment (Manabe et al., 2007).

From these observations, it becomes obvious that Imazapyr-treated rice seedlings cope with the impaired branched-chain amino acid synthesis, first through increased protein turnover (Royuela et al., 2000). However, as this results in growth disturbance, carbon is redirected to non-growth associated macromolecules such as free sugars, starch and free fatty acids (Zabalza et al., 2004). In addition to providing a flexible storage pool for the assimilated carbon, synthesis of these components helps the plant get rid of excess ATP and redox power, caused by impaired growth. Through ^{13}C-INST-MFA, previous macro-compositional observations could be related to Imazapyr-treatment on the molecular level. In this way, the metabolic story of Imazapyr-treatment could be unravelled on a deeper level from the intracellular fluxes. Here, it becomes obvious that Imazapyr not only effects a single enzyme, but influences cellular metabolism on the systems-level. In this regard, ^{13}C-INST-MFA in plants promises many applications, e.g. in the elucidation of mode of action, of the effect of genetic modifications as well as of environmental stress response.

7.   Conclusion and Outlook

In the present work, both *in silico* and *in vivo* methods for flux analysis in plants were successfully developed and applied. A highly complex network model of *Arabidopsis thaliana* allowed realistic integration of *in silico* fluxes with *in vivo* data, which is, so far, one of the few examples hereof. Furthermore, the first *in vivo* flux analysis of an agriculturally relevant crop and its first real-life case-study in whole plants are presented. Both analyses enhance understanding of plant physiology and have great potential in green biotechnology, as these concepts can now also be adopted to other crops, plant lines and stress inducers, such as abiotic stresses, herbicides and fungicides.

In addition to improving our understanding of plant metabolism, generated fluxes can be analyzed in detail to guide genetic engineering of superior plant lines (Fig. 7-1). Genetic targets can be predicted by comparing fluxes under different conditions or in specific mutants, or by performing *in silico* analysis based on the flux fingerprint. In this way, rational metabolic engineering could assist in establishing plants as green factories for industrial applications (Yuan & Grötewold, 2015).

Future investigation of plant metabolism through metabolic modeling would particularly benefit from increased throughput and resolution. Throughput-improvements of the labeling workflow can be accomplished by simultaneous labeling of multiple plants, e.g. directly in the greenhouse, which could be combined with an automated sampling procedure by robots, similar to those applied in high-throughput automated phenotypic screening (Junker et al., 2014). Accelerated analytical processing can be achieved by developing completely autonomous analytical pipelines (Ngounou Wetie et al., 2013; Núñez et al., 2013; Shubhakar et al., 2015). In addition, computational tools, such as our own MATLAB-based software tool, allow for automated data-processing of large-scale mass spectrometry data from isotopically instationary labeling

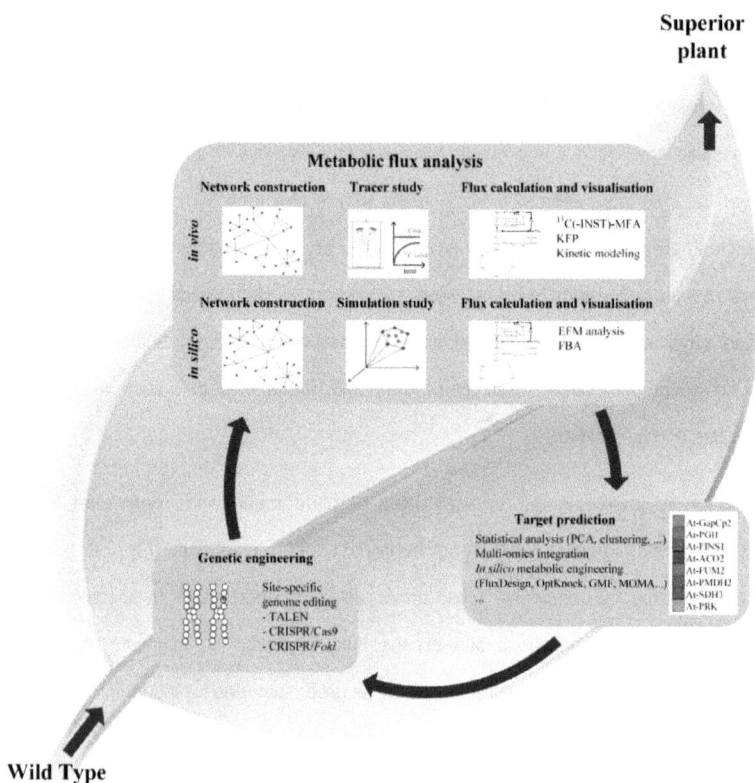

Figure 7-1: Iterative development of flux-analysis-based plant engineering

After selecting the plant system best suited for the pursued goal, experimental data is acquired. The experimental effort greatly depends on the chosen flux analysis approach and can range from merely stoichiometric model construction to elaborative ^{13}C-based profiling. Flux analysis is particularly useful in identifying possible targets for subsequent metabolic/genetic engineering and thus can be applied to create plants that are useful as industrial green factories.

experiments, including correction for natural isotopic abundance and data-preparation for selected flux estimation software. Furthermore, enhancing efficiency by diminishing calculation time of flux estimation has already been realized by reducing the number of isotopomeric units that are considered, based on elementary metabolite units (EMUs) (Antoniewicz et al., 2007; Young et al., 2008). Newly emerging concepts in this field, include the use of fluxomers (Srour et al., 2011) and the combination of EMUs with flux coupling to achieve faster algorithms (Suthers et al., 2010). Also, the development of multi-processor systems and computing clusters increase throughput significantly (Junker, 2014). *In silico* analysis of plant metabolism would benefit from future technological development, directed at providing faster methods, able to process increasingly larger networks, as structural pathway analysis is confronted with impaired scalability to genome-wide models.

The ultimate goal of whole plant flux analysis would be to simultaneously model multiple interconnected organs, as this mimics natural conditions most realistically. However, here we face the problem of scalability and it is to be awaited how much more data input is needed to sufficiently restrain such flux estimations. In this context, combining existing methods is promising, as shown for a combination of FBA and kinetic modeling to investigate multi-tissue systems (Grafahrend-Belau et al., 2013) and the consolidation of kinetic modeling with advection-diffusion modeling in the phloem (Rohwer, 2012). In this regard, the integration of FBA and $^{13}C(-INST)-MFA$ is conceivable to elucidate the plant fluxome.

8.   Symbols and Abbreviations

Symbols

α	significance level
$\alpha_{i,obj}$	target potential coefficient of reaction i under a certain objective function
δ	standard deviation
Δc	change in pool sizes
Δp	change in parameter vector
$\Delta \Phi$	change in SSR value
Δv	change in flux vector
g	gradient vector
J	Jacobian matrix
H	Hessian matrix
K_m and K_i	enzyme specific constants
λ	damping parameter
ξ	molar C-content
r	rate measurements
r_m	experimentally determined rate measurements
R_i	reaction rate
σ	variance
s	stoichiometric factor
S_i	metabolite concentration
S_{mxn}	stoichiometric matrix with m metabolites and n reactions
t	time
Φ	objective function
$\chi^2(n-p)$	chi-square distribution with n-p degrees of freedom
v	flux vector

v_{max}	maximal enzymatic activity
x	labeling states
x_m	measured labeling states
$Y_{P/C}$	product yield on substrate C

Abbreviations

[13]C-INST-MFA	^{13}C-labeling based isotopically instationary metabolic flux analysis
[13]C-MFA	^{13}C-labeling based metabolic flux analysis
ADP	adenosine diphosphate
AHAS	acetohydroxyacid synthase
ATP	adenosine triphosphate
BMY	biomass yield
CBB cycle	Calvin-Benson-Bassham cycle
CE-MS	capillary-electrophoresis coupled to mass spectrometry
CoA	Coenzyme A
DMSO	dimethyl sulfoxide
DNA	deoxyribonucleic acid
DOF	degrees of freedom
DW	dry weight
EA-C-irMS	elemental analyzer combustion isotope ratio mass spectrometer
EFM	elementary flux mode
EFMA	elementary flux mode analysis
EMP pathway	Embden-Meyerhof-Parnas pathway
EMU	elementary metabolite unit
EPDM	ethylene propylene diene monomer
FAME	fatty acid methyl ester
FBA	flux balance analysis
GC-C-irMS	gas chromatrography combustion isotope ratio mass spectrometry
GC-irMS	gas chromatrography-isotope ratio mass spectrometry
GC-MS	gas chromatography-mass spectrometry
GUI	graphical user interface
INCA	Isotopomer Network Compartmental Analysis

IPPP	iso-pentenyl pyrophosphate
ISTD	internal standard
KFP	kinetic flux profiling
LC-MS/MS	liquid chromatography-mass spectrometry/mass spectrometry
L-Ma	Levenberg-Marquardt algorithm
Mb	megabases
MCA	Monte-Carlo Analysis
MID	mass isotopomer distribution
MFA	metabolic flux analysis
MS	mass spectrometry
MSD	mass selective detector
NAD(P)	nicotinamide adenine dinucleotide (phosphate), oxidized
NAD(P)H	nicotinamide adenine dinucleotide (phosphate), reduced
NMR	nuclear magnetic resonance
MTBE	methyl *tert*-butyl ether
ODE	ordinary differential equation
PAR	photosynthetically active radiation
PP pathway	pentose phosphate pathway
ParCon	Parameter Continuation
QP	quadratic programming
RAM	random access memory
RGR	relative growth rate
RuBisCO	ribulose-1,5-bisphosphate carboxylase/oxygenase
SSR	residual sum of squares
TCA cycle	tricarboxylic acid cycle
TMSA	trimethylsulfonium hydroxide

9.   Bibliography

Acuña, V., Chierichetti, F., Lacroix, V., Marchetti-Spaccamela, A., Sagot, M. F. and Stougie, L. (2009). Modes and cuts in metabolic networks: complexity and algorithms. *Biosystems, 95*(1), 51-60. doi: 10.1016/j.biosystems.2008.06.015

Ajikumar, P. K., Xiao, W. H., Tyo, K. E., Wang, Y., Simeon, F., Leonard, E., Mucha, O., Phon, T. H., Pfeifer, B. and Stephanopoulos, G. (2010). Isoprenoid pathway optimization for Taxol precursor overproduction in *Escherichia coli. Science, 330*(6000), 70-74. doi: 10.1126/science.1191652

Allen, J. F. (2003). Cyclic, pseudocyclic and noncyclic photophosphorylation: new links in the chain. *Trends Plant Sci, 8*(1), 15-19.

Alonso, A. P., Goffman, F. D., Ohlrogge, J. B. and Shachar-Hill, Y. (2007). Carbon conversion efficiency and central metabolic fluxes in developing sunflower (*Helianthus annuus* L.) embryos. *Plant J, 52*(2), 296-308. doi: 10.1111/j.1365-313X.2007.03235.x

Alonso, A. P., Vigeolas, H., Raymond, P., Rolin, D. and Dieuaide-Noubhani, M. (2005). A new substrate cycle in plants. Evidence for a high glucose-phosphate-to-glucose turnover from *in vivo* steady-state and pulse-labeling experiments with [^{13}C]glucose and [^{14}C]glucose. *Plant Physiol, 138*(4), 2220-2232. doi: 10.1104/pp.105.062083

Amir, R. (2010). Current understanding of the factors regulating methionine content in vegetative tissues of higher plants. *Amino Acids, 39*(4), 917-931. doi: 10.1007/s00726-010-0482-x

Anai, T., Koga, M., Tanaka, H., Kinoshita, T., Rahman, S. M. and Takagi, Y. (2003). Improvement of rice (*Oryza sativa* L.) seed oil quality through introduction of a soybean microsomal omega-3 fatty acid desaturase gene. *Plant Cell Rep, 21*(10), 988-992. doi: 10.1007/s00299-003-0609-6

Andersen, A., Sorensen, H. and Nielsen, G. (1961). Growth Chamber for Labelling Plant Material UniformLy with Radiocarbon. *Physiologia Plantarum, 14*(2), 378-&. doi: DOI 10.1111/j.1399-3054.1961.tb07872.x

Antoniewicz, M. R. (2015). Methods and advances in metabolic flux analysis: a mini-review. *J Ind Microbiol Biotechnol, 42*(3), 317-325. doi: 10.1007/s10295-015-1585-x

Antoniewicz, M. R., Kelleher, J. K. and Stephanopoulos, G. (2006). Determination of confidence intervals of metabolic fluxes estimated from stable isotope measurements. *Metab Eng, 8*(4), 324-337. doi: 10.1016/j.ymben.2006.01.004

Antoniewicz, M. R., Kelleher, J. K. and Stephanopoulos, G. (2007). Elementary metabolite units (EMU): a novel framework for modeling isotopic distributions. *Metab Eng, 9*(1), 68-86. doi: 10.1016/j.ymben.2006.09.001

Arabidopsis Genome Initiative. (2000). Analysis of the genome sequence of the flowering plant *Arabidopsis thaliana. Nature, 408*(6814), 796-815. doi: 10.1038/35048692

AraCyc. (release 8.0, 2014). from http://pmn.plantcyc.org

Araújo, W. L., Nunes-Nesi, A. and Fernie, A. R. (2014). On the role of plant mitochondrial metabolism and its impact on photosynthesis in both optimal and sub-optimal growth conditions. *Photosynth Res, 119*(1-2), 141-156. doi: 10.1007/s11120-013-9807-4

Arnqvist, L., Persson, M., Jonsson, L., Dutta, P. C. and Sitbon, F. (2008). Overexpression of CYP710A1 and CYP710A4 in transgenic *Arabidopsis* plants increases the level of stigmasterol at the expense of sitosterol. *Planta, 227*(2), 309-317. doi: 10.1007/s00425-007-0618-8

Assmus, H. E. (2005). *Modelling the carbohydrate metabolism in potato tuber cells.* (PhD thesis), Univ. Oxford, Oxford Brookes

Baerenfaller, K., Massonnet, C., Walsh, S., Baginsky, S., Buhlmann, P., Hennig, L., Hirsch-Hoffmann, M., Howell, K. A., Kahlau, S., Radziejwoski, A., Russenberger, D., Rutishauser, D., Small, I., Stekhoven, D., Sulpice, R., Svozil, J., Wuyts, N., Stitt, M., Hilson, P., Granier, C. and Gruissem, W. (2012). Systems-based analysis of *Arabidopsis* leaf growth reveals adaptation to water deficit. *Mol Syst Biol, 8*, 606. doi: 10.1038/msb.2012.39

Bajaj, S. and Mohanty, A. (2005). Recent advances in rice biotechnology--towards genetically superior transgenic rice. *Plant Biotechnol J, 3*(3), 275-307. doi: 10.1111/j.1467-7652.2005.00130.x

Balcke, G. U., Kolle, S. N., Kamp, H., Bethan, B., Looser, R., Wagner, S., Landsiedel, R. and van Ravenzwaay, B. (2011). Linking energy metabolism to dysfunctions in mitochondrial respiration--a metabolomics *in vitro* approach. *Toxicol Lett, 203*(3), 200-209. doi: 10.1016/j.toxlet.2011.03.013

Baroja-Fernández, E., Muñoz, F. J., Akazawa, T. and Pozueta-Romero, J. (2001). Reappraisal of the currently prevailing model of starch biosynthesis in photosynthetic tissues: a proposal involving the cytosolic production of ADP-glucose by sucrose synthase and occurrence of cyclic turnover of starch in the chloroplast. *Plant Cell Physiol, 42*(12), 1311-1320.

Becker, J., Reinefeld, J., Stellmacher, R., Schäfer, R., Lange, A., Meyer, H., Lalk, M., Zelder, O., von Abendroth, G., Schröder, H., Haefner, S. and Wittmann, C. (2013). Systems-wide analysis and engineering of metabolic pathway fluxes in bio-succinate producing *Basfia succiniciproducens*. *Biotechnol Bioeng, 110*(11), 3013-3023. doi: 10.1002/bit.24963

Becker, J., Zelder, O., Häfner, S., Schröder, H. and Wittmann, C. (2011). From zero to hero--design-based systems metabolic engineering of *Corynebacterium glutamicum* for L-lysine production. *Metab Eng, 13*(2), 159-168. doi: 10.1016/j.ymben.2011.01.003

Berg, J. M., Tymoczko, J. L. and Stryer, L. (2007). *Biochemistry* (6th ed.): W.H. Freeman and Company.

Beurton-Aimar, M., Beauvoit, B., Monier, A., Vallée, F., Dieuaide-Noubhani, M. and Colombié, S. (2011). Comparison between elementary flux modes analysis and 13C-metabolic fluxes measured in bacterial and plant cells. *BMC Syst Biol, 5*, 95. doi: 10.1186/1752-0509-5-95

Bhullar, N. K. and Gruissem, W. (2013). Nutritional enhancement of rice for human health: the contribution of biotechnology. *Biotechnol Adv, 31*(1), 50-57. doi: 10.1016/j.biotechadv.2012.02.001

Blank, L. M., Kuepfer, L. and Sauer, U. (2005). Large-scale 13C-flux analysis reveals mechanistic principles of metabolic network robustness to null mutations in yeast. *Genome Biol, 6*(6), R49. doi: 10.1186/gb-2005-6-6-r49

Bläsing, O. E., Gibon, Y., Günther, M., Höhne, M., Morcuende, R., Osuna, D., Thimm, O., Usadel, B., Scheible, W. R. and Stitt, M. (2005). Sugars and circadian regulation make major contributions to the global regulation of diurnal gene expression in *Arabidopsis*. *Plant Cell, 17*(12), 3257-3281. doi: 10.1105/tpc.105.035261

Buchanan, B. B. and Balmer, Y. (2005). Redox regulation: a broadening horizon. *Annu Rev Plant Biol, 56*, 187-220. doi: 10.1146/annurev.arplant.56.032604.144246

Campo, S., Baldrich, P., Messeguer, J., Lalanne, E., Coca, M. and San Segundo, B. (2014). Overexpression of a Calcium-Dependent Protein Kinase Confers Salt and Drought Tolerance in Rice by Preventing Membrane Lipid Peroxidation. *Plant Physiol, 165*(2), 688-704. doi: 10.1104/pp.113.230268

Chalhoub, B., Denoeud, F., Liu, S., Parkin, I. A., Tang, H., Wang, X., Chiquet, J., Belcram, H., Tong, C., Samans, B., Correa, M., Da Silva, C., Just, J., Falentin, C., Koh, C. S., Le Clainche, I., Bernard, M., Bento, P., Noel, B., Labadie, K., Alberti, A., Charles, M., Arnaud, D., Guo, H., Daviaud, C., Alamery, S., Jabbari, K., Zhao, M., Edger, P. P., Chelaifa, H., Tack, D., Lassalle, G., Mestiri, I., Schnel, N., Le Paslier, M. C., Fan, G., Renault, V., Bayer, P. E., Golicz, A. A., Manoli, S., Lee, T. H., Thi, V. H., Chalabi, S., Hu, Q., Fan, C., Tollenaere, R., Lu, Y., Battail, C., Shen, J., Sidebottom, C. H., Wang, X., Canaguier, A., Chauveau, A., Bérard, A., Deniot, G., Guan, M., Liu, Z., Sun, F., Lim, Y. P., Lyons, E., Town, C. D., Bancroft, I., Wang, X., Meng, J., Ma, J., Pires, J. C., King, G. J., Brunel, D., Delourme, R., Renard, M., Aury, J. M., Adams, K. L., Batley, J., Snowdon, R. J., Tost, J., Edwards, D., Zhou, Y., Hua, W., Sharpe, A. G., Paterson, A. H., Guan, C. and Wincker, P. (2014). Plant genetics. Early allopolyploid evolution in the post-Neolithic *Brassica napus* oilseed genome. *Science, 345*(6199), 950-953. doi: 10.1126/science.1253435

Chang, A. K. and Duggleby, R. G. (1997). Expression, purification and characterization of Arabidopsis thaliana acetohydroxyacid synthase. *Biochem J, 327 (Pt 1)*, 161-169.

Chang, R. L., Ghamsari, L., Manichaikul, A., Hom, E. F., Balaji, S., Fu, W., Shen, Y., Hao, T., Palsson, B. Ø., Salehi-Ashtiani, K. and Papin, J. A. (2011). Metabolic network reconstruction of *Chlamydomonas* offers insight into light-driven algal metabolism. *Mol Syst Biol, 7*, 518. doi: 10.1038/msb.2011.52

Chao, J. F., Quick, W. A., Hsiao, A. I. and Xie, H. S. (1994). Effect of Imazamethabenz on Histology and Histochemistry of Polysaccharides in the Main Shoot of Wild Oat (Avena-Fatua). *Weed Science, 42*(3), 345-352.

Chao, Y. Y., Hong, C. Y. and Kao, C. H. (2010). The decline in ascorbic acid content is associated with cadmium toxicity of rice seedlings. *Plant Physiol Biochem, 48*(5), 374-381. doi: 10.1016/j.plaphy.2010.01.009

Chastain, C. J., Fries, J. P., Vogel, J. A., Randklev, C. L., Vossen, A. P., Dittmer, S. K., Watkins, E. E., Fiedler, L. J., Wacker, S. A., Meinhover, K. C., Sarath, G. and Chollet, R. (2002). Pyruvate,orthophosphate dikinase in leaves and chloroplasts of C_3 plants undergoes light-/dark-induced reversible phosphorylation. *Plant Physiol, 128*(4), 1368-1378. doi: 10.1104/pp.010806

Chen, M., Wei, H., Cao, J., Liu, R., Wang, Y. and Zheng, C. (2007). Expression of *Bacillus subtilis* proBA genes and reduction of feedback inhibition of proline synthesis increases proline production and confers osmotolerance in transgenic *Arabidopsis*. *J Biochem Mol Biol, 40*(3), 396-403.

Cheung, C. Y., Williams, T. C., Poolman, M. G., Fell, D. A., Ratcliffe, R. G. and Sweetlove, L. J. (2013). A method for accounting for maintenance costs in flux balance analysis improves the prediction of plant cell metabolic phenotypes under stress conditions. *Plant J, 75*(6), 1050-1061. doi: 10.1111/tpj.12252

Chujo, T., Miyamoto, K., Ogawa, S., Masuda, Y., Shimizu, T., Kishi-Kaboshi, M., Takahashi, A., Nishizawa, Y., Minami, E., Nojiri, H., Yamane, H. and Okada, K. (2014). Overexpression of phosphomimic mutated *OsWRKY53* leads to enhanced blast resistance in rice. *PLoS One, 9*(6), e98737. doi: 10.1371/journal.pone.0098737

Coplen, T. B., Brand, W. A., Gehre, M., Groning, M., Meijer, H. A., Toman, B., Verkouteren, R. M. and International Atomic Energy, A. (2006). After two decades a second anchor for the VPDB delta^{13}C scale. *Rapid Commun Mass Spectrom, 20*(21), 3165-3166. doi: 10.1002/rcm.2727

Cosgrove, D. J. (2005). Growth of the plant cell wall. *Nat Rev Mol Cell Biol, 6*(11), 850-861. doi: 10.1038/nrm1746

Costenoble, R., Picotti, P., Reiter, L., Stallmach, R., Heinemann, M., Sauer, U. and Aebersold, R. (2011). Comprehensive quantitative analysis of central carbon and amino-acid metabolism in *Saccharomyces cerevisiae* under multiple conditions by targeted proteomics. *Mol Syst Biol, 7*, 464. doi: 10.1038/msb.2010.122

Coudert, Y., Périn, C., Courtois, B., Khong, N. G. and Gantet, P. (2010). Genetic control of root development in rice, the model cereal. *Trends Plant Sci, 15*(4), 219-226. doi: 10.1016/j.tplants.2010.01.008

Cruz, J. A., Avenson, T. J., Kanazawa, A., Takizawa, K., Edwards, G. E. and Kramer, D. M. (2005). Plasticity in light reactions of photosynthesis for energy production and photoprotection. *J Exp Bot, 56*(411), 395-406. doi: 10.1093/jxb/eri022

Cusido, R. M., Onrubia, M., Sabater-Jara, A. B., Moyano, E., Bonfill, M., Goossens, A., Angeles Pedreño, M. and Palazon, J. (2014). A rational approach to improving the biotechnological production of taxanes in plant cell cultures of *Taxus spp. Biotechnol Adv, 32*(6), 1157-1167. doi: 10.1016/j.biotechadv.2014.03.002

de Bossoreille de Ribou, S., Douam, F., Hamant, O., Frohlich, M. W. and Negrutiu, I. (2013). Plant science and agricultural productivity: why are we hitting the yield ceiling? *Plant Sci, 210*, 159-176. doi: 10.1016/j.plantsci.2013.05.010

de Oliveira Dal'Molin, C. G. and Nielsen, L. K. (2013). Plant genome-scale metabolic reconstruction and modelling. *Curr Opin Biotechnol, 24*(2), 271-277. doi: 10.1016/j.copbio.2012.08.007

de Oliveira Dal'Molin, C. G., Quek, L. E., Palfreyman, R. W., Brumbley, S. M. and Nielsen, L. K. (2010a). AraGEM, a genome-scale reconstruction of the primary metabolic network in Arabidopsis. *Plant Physiol, 152*(2), 579-589. doi: 10.1104/pp.109.148817

de Oliveira Dal'Molin, C. G., Quek, L. E., Palfreyman, R. W., Brumbley, S. M. and Nielsen, L. K. (2010b). C4GEM, a genome-scale metabolic model to study C4 plant metabolism. *Plant Physiol, 154*(4), 1871-1885. doi: 10.1104/pp.110.166488

de Oliveira Dal'Molin, C. G., Quek, L. E., Palfreyman, R. W. and Nielsen, L. K. (2011). AlgaGEM--a genome-scale metabolic reconstruction of algae based on the *Chlamydomonas reinhardtii* genome. *BMC Genomics, 12 Suppl 4*, S5. doi: 10.1186/1471-2164-12-S4-S5

de Oliveira Dal'Molin, C. G., Quek, L. E., Saa, P. A. and Nielsen, L. K. (2015). A multi-tissue genome-scale metabolic modeling framework for the analysis of whole plant systems. *Front Plant Sci,* 6, 4. doi: 10.3389/fpls.2015.00004

Dennis, J. E. and Schnabel, R. B. (1983). *Numerical methods for unconstrained optimization and nonlinear equations.* Englewood Cliffs, N.J., USA: Prentice-Hall.

Dersch, L. (2016). *Metabolic fluxes of the model crop rice: Technological development and biological applications.* (PhD thesis), Saarland University, Germany.

Dersch, L., Beckers, V., Rasch, D., Melzer, G., Bolten, C. J., Kiep, K., Becker, H., Bläsing, O. E., Fuchs, R., Ehrhardt, T. and Wittmann, C. (2016a). *In Vivo* Assimilation, Translocation and Molecular Carbon and Nitrogen Fluxes in the Crop *Oryza sativa. Plant Physiol, accepted.*

Dersch, L. M., Beckers, V. and Wittmann, C. (2016b). Green Pathways: Metabolic network analysis of plant systems. *Metab Eng,* 34, 1-24. doi: 10.1016/j.ymben.2015.12.001.

Dey, P. M. and Harborn, J. B. (1997). *Plant Biochemistry.* London: Academic Press.

Driouch, H., Melzer, G. and Wittmann, C. (2012). Integration of *in vivo* and *in silico* metabolic fluxes for improvement of recombinant protein production. *Metab Eng, 14*(1), 47-58. doi: 10.1016/j.ymben.2011.11.002

Duggleby, R. G., McCourt, J. A. and Guddat, L. W. (2008). Structure and mechanism of inhibition of plant acetohydroxyacid synthase. *Plant Physiol Biochem, 46*(3), 309-324. doi: 10.1016/j.plaphy.2007.12.004

Dyer, J. M., Stymne, S., Green, A. G. and Carlsson, A. S. (2008). High-value oils from plants. *Plant J, 54*(4), 640-655. doi: 10.1111/j.1365-313X.2008.03430.x

Ehleringer, J. and Bjorkman, O. (1977). Quantum Yields for CO_2 Uptake in C_3 and C_4 Plants: Dependence on Temperature, CO_2, and O_2 Concentration. *Plant Physiol, 59*(1), 86-90.

Ehleringer, J. and Pearcy, R. W. (1983). Variation in Quantum Yield for CO_2 Uptake among C_3 and C_4 Plants. *Plant Physiol, 73*(3), 555-559.

Eicks, M., Maurino, V., Knappe, S., Flugge, U. I. and Fischer, K. (2002). The plastidic pentose phosphate translocator represents a link between the cytosolic and the plastidic pentose phosphate pathways in plants. *Plant Physiol, 128*(2), 512-522. doi: 10.1104/pp.010576

Facchinelli, F. and Weber, A. P. (2011). The metabolite transporters of the plastid envelope: an update. *Front Plant Sci, 2,* 50. doi: 10.3389/fpls.2011.00050

Fan, J., Yan, C., Zhang, X. and Xu, C. (2013). Dual role for phospholipid:diacylglycerol acyltransferase: enhancing fatty acid synthesis and diverting fatty acids from membrane lipids to triacylglycerol in *Arabidopsis* leaves. *Plant Cell, 25*(9), 3506-3518. doi: 10.1105/tpc.113.117358

FAO. (2013). FAOSTAT Food and Agricultural Organization of the United Nations, Rome, Italy. Retrieved 15.04.2015, from www.faostat.org

Fatland, B. L., Nikolau, B. J. and Wurtele, E. S. (2005). Reverse genetic characterization of cytosolic acetyl-CoA generation by ATP-citrate lyase in *Arabidopsis. Plant Cell, 17*(1), 182-203. doi: 10.1105/tpc.104.026211

Fernie, A. R., Carrari, F. and Sweetlove, L. J. (2004). Respiratory metabolism: glycolysis, the TCA cycle and mitochondrial electron transport. *Curr Opin Plant Biol, 7*(3), 254-261. doi: 10.1016/j.pbi.2004.03.007

Fernie, A. R. and Morgan, J. A. (2013). Analysis of metabolic flux using dynamic labelling and metabolic modelling. *Plant Cell Environ, 36*(9), 1738-1750. doi: 10.1111/pce.12083

Fischer, K. (2011). The import and export business in plastids: transport processes across the inner envelope membrane. *Plant Physiol, 155*(4), 1511-1519. doi: 10.1104/pp.110.170241

Fischer, W. N., André, B., Rentsch, D., Krolkiewicz, S., Tegeder, M., Breitkreuz, K. and Frommer, W. B. (1998). Amino acid transport in plants. *Trends Plant Sci, 3*(5), 188-195. doi: Doi 10.1016/S1360-1385(98)01231-X

Foyer, C. H., Neukermans, J., Queval, G., Noctor, G. and Harbinson, J. (2012). Photosynthetic control of electron transport and the regulation of gene expression. *J Exp Bot, 63*(4), 1637-1661. doi: 10.1093/jxb/ers013

Fuchs, R., Woiwode, O., Peter, E., Schön, H., Wittmann, C., Rasch, D., Beckers, V. and Dersch, L. (2014) Method and device for marking isotopes. WO2014079696 A1.

Furumoto, T., Yamaguchi, T., Ohshima-Ichie, Y., Nakamura, M., Tsuchida-Iwata, Y., Shimamura, M., Ohnishi, J., Hata, S., Gowik, U., Westhoff, P., Bräutigam, A., Weber, A. P. and Izui, K. (2011). A plastidial sodium-dependent pyruvate transporter. *Nature, 476*(7361), 472-475. doi: 10.1038/nature10250

Garris, A. J., Tai, T. H., Coburn, J., Kresovich, S. and McCouch, S. (2005). Genetic structure and diversity in *Oryza sativa* L. *Genetics, 169*(3), 1631-1638. doi: 10.1534/genetics.104.035642

Ghosh, A., Pareek, A., Sopory, S. K. and Singla-Pareek, S. L. (2014). A glutathione responsive rice glyoxalase II, OsGLYII-2, functions in salinity adaptation by maintaining better photosynthesis efficiency and anti-oxidant pool. *Plant J, 80*(1), 93-105. doi: 10.1111/tpj.12621

Gibson, K., Park, J. S., Nagai, Y., Hwang, S. K., Cho, Y. C., Roh, K. H., Lee, S. M., Kim, D. H., Choi, S. B., Ito, H., Edwards, G. E. and Okita, T. W. (2011). Exploiting leaf starch synthesis as a transient sink to elevate photosynthesis, plant productivity and yields. *Plant Sci, 181*(3), 275-281. doi: 10.1016/j.plantsci.2011.06.001

Goff, S. A. (1999). Rice as a model for cereal genomics. *Curr Opin Plant Biol, 2*(2), 86-89. doi: 10.1016/S1369-5266(99)80018-1

Goffman, F. D., Alonso, A. P., Schwender, J., Shachar-Hill, Y. and Ohlrogge, J. B. (2005). Light enables a very high efficiency of carbon storage in developing embryos of rapeseed. *Plant Physiol, 138*(4), 2269-2279. doi: 10.1104/pp.105.063628

Goto, S., Sasakura-Shimoda, F., Suetsugu, M., Selvaraj, M. G., Hayashi, N., Yamazaki, M., Ishitani, M., Shimono, M., Sugano, S., Matsushita, A., Tanabata, T. and Takatsuji, H. (2014). Development of disease-resistant rice by optimized expression of *WRKY45*. *Plant Biotechnol J*. doi: 10.1111/pbi.12303

Grafahrend-Belau, E., Junker, A., Eschenröder, A., Müller, J., Schreiber, F. and Junker, B. H. (2013). Multiscale metabolic modeling: dynamic flux balance analysis on a whole-plant scale. *Plant Physiol, 163*(2), 637-647. doi: 10.1104/pp.113.224006

Grafahrend-Belau, E., Schreiber, F., Koschützki, D. and Junker, B. H. (2009). Flux balance analysis of barley seeds: a computational approach to study systemic properties of central metabolism. *Plant Physiol, 149*(1), 585-598. doi: 10.1104/pp.108.129635

Grafahrend-Belau, E., Weise, S., Koschutzki, D., Scholz, U., Junker, B. H. and Schreiber, F. (2008). MetaCrop: a detailed database of crop plant metabolism. *Nucleic Acids Res, 36*(Database issue), D954-958. doi: 10.1093/nar/gkm835

Gramene. A comparative resource for plants. Retrieved 15.04.2015, from http://ensembl.gramene.org/Zea_mays/Info/Index?db=core

Haferkamp, I., Fernie, A. R. and Neuhaus, H. E. (2011). Adenine nucleotide transport in plants: much more than a mitochondrial issue. *Trends Plant Sci, 16*(9), 507-515. doi: 10.1016/j.tplants.2011.04.001

Hall, D. O. and Rao, K. (1999). *Photosynthesis*. Cambridge University Press.

Hashimoto, H., Kurahotta, M. and Katoh, S. (1989). Changes in Protein-Content and in the Structure and Number of Chloroplasts during Leaf Senescence in Rice Seedlings. *Plant and Cell Physiology, 30*(5), 707-715.

Hashimoto, M. and Komatsu, S. (2007). Proteomic analysis of rice seedlings during cold stress. *Proteomics, 7*(8), 1293-1302. doi: 10.1002/pmic.200600921

Hasunuma, T., Harada, K., Miyazawa, S., Kondo, A., Fukusaki, E. and Miyake, C. (2010). Metabolic turnover analysis by a combination of *in vivo* ^{13}C-labelling from [13]CO2 and metabolic profiling with CE-MS/MS reveals rate-limiting steps of the C3 photosynthetic pathway in *Nicotiana tabacum* leaves. *J Exp Bot, 61*(4), 1041-1051. doi: 10.1093/jxb/erp374

Hasunuma, T., Okazaki, F., Okai, N., Hara, K. Y., Ishii, J. and Kondo, A. (2013). A review of enzymes and microbes for lignocellulosic biorefinery and the possibility of their application to consolidated bioprocessing technology. *Bioresour Technol, 135*, 513-522. doi: 10.1016/j.biortech.2012.10.047

Hay, J. O., Shi, H., Heinzel, N., Hebbelmann, I., Rolletschek, H. and Schwender, J. (2014). Integration of a constraint-based metabolic model of *Brassica napus* developing seeds with ^{13}C-metabolic flux analysis. *Front Plant Sci, 5*, 724. doi: 10.3389/fpls.2014.00724

Heinzle, E., Yuan, Y., Kumar, S., Wittmann, C., Gehre, M., Richnow, H. H., Wehrung, P., Adam, P. and Albrecht, P. (2008). Analysis of ^{13}C labeling enrichment in microbial culture applying metabolic tracer experiments using gas chromatography-combustion-isotope ratio mass spectrometry. *Anal Biochem, 380*(2), 202-210. doi: 10.1016/j.ab.2008.05.039

Herrero, J., Fernández-Pérez, F., Yebra, T., Novo-Uzal, E., Pomar, F., Pedreño, M. A., Cuello, J., Guéra, A., Esteban-Carrasco, A. and Zapata, J. M. (2013). Bioinformatic and functional characterization of the basic peroxidase 72 from *Arabidopsis thaliana* involved in lignin biosynthesis. *Planta, 237*(6), 1599-1612. doi: 10.1007/s00425-013-1865-5

High, S. M., Cohen, M. B., Shu, Q. Y. and Altosaar, I. (2004). Achieving successful deployment of *Bt* rice. *Trends Plant Sci, 9*(6), 286-292. doi: 10.1016/j.tplants.2004.04.002

Hoefnagel, M. H. N., Atkin, O. K. and Wiskich, J. T. (1998). Interdependence between chloroplasts and mitochondria in the light and the dark. *Biochimica Et Biophysica Acta-Bioenergetics, 1366*(3), 235-255. doi: Doi 10.1016/S0005-2728(98)00126-1

Hu, J., Baker, A., Bartel, B., Linka, N., Mullen, R. T., Reumann, S. and Zolman, B. K. (2012). Plant peroxisomes: biogenesis and function. *Plant Cell, 24*(6), 2279-2303. doi: 10.1105/tpc.112.096586

Huege, J., Sulpice, R., Gibon, Y., Lisec, J., Koehl, K. and Kopka, J. (2007). GC-EI-TOF-MS analysis of *in vivo* carbon-partitioning into soluble metabolite pools of higher plants by monitoring isotope dilution after $^{13}CO_2$ labelling. *Phytochemistry, 68*(16-18), 2258-2272. doi: 10.1016/j.phytochem.2007.03.026

Humphreys, J. M. and Chapple, C. (2002). Rewriting the lignin roadmap. *Curr Opin Plant Biol, 5*(3), 224-229.

Hurkman, W. J. and Tanaka, C. K. (1986). Solubilization of plant membrane proteins for analysis by two-dimensional gel electrophoresis. *Plant Physiol, 81*(3), 802-806.

Hwang, K. S., Kim, H. U., Charusanti, P., Palsson, B. Ø. and Lee, S. Y. (2014). Systems biology and biotechnology of *Streptomyces* species for the production of secondary metabolites. *Biotechnol Adv, 32*(2), 255-268. doi: 10.1016/j.biotechadv.2013.10.008

International Barley Genome Sequencing, C., Mayer, K. F., Waugh, R., Brown, J. W., Schulman, A., Langridge, P., Platzer, M., Fincher, G. B., Muehlbauer, G. J., Sato, K., Close, T. J., Wise, R. P. and Stein, N. (2012). A physical, genetic and functional sequence assembly of the barley genome. *Nature, 491*(7426), 711-716. doi: 10.1038/nature11543

International Rice Genome Sequencing Project. (2005). The map-based sequence of the rice genome. *Nature, 436*(7052), 793-800. doi: 10.1038/nature03895

Ishihara, A., Hashimoto, Y., Tanaka, C., Dubouzet, J. G., Nakao, T., Matsuda, F., Nishioka, T., Miyagawa, H. and Wakasa, K. (2008). The tryptophan pathway is involved in the defense responses of rice against pathogenic infection via serotonin production. *Plant J, 54*(3), 481-495. doi: 10.1111/j.1365-313X.2008.03441.x

Itoh, J., Nonomura, K., Ikeda, K., Yamaki, S., Inukai, Y., Yamagishi, H., Kitano, H. and Nagato, Y. (2005). Rice plant development: from zygote to spikelet. *Plant Cell Physiol, 46*(1), 23-47. doi: 10.1093/pcp/pci501

Jander, G. and Joshi, V. (2009). Aspartate-Derived Amino Acid Biosynthesis in *Arabidopsis thaliana*. *Arabidopsis Book, 7*, e0121. doi: 10.1199/tab.0121

Jensen, M. V., Joseph, J. W., Ronnebaum, S. M., Burgess, S. C., Sherry, A. D. and Newgard, C. B. (2008). Metabolic cycling in control of glucose-stimulated insulin secretion. *Am J Physiol Endocrinol Metab, 295*(6), E1287-1297. doi: 10.1152/ajpendo.90604.2008

Jensen, R. G. (2000). Activation of Rubisco regulates photosynthesis at high temperature and CO_2. *Proc Natl Acad Sci U S A, 97*(24), 12937-12938. doi: 10.1073/pnas.97.24.12937

Jeong, J. S., Kim, Y. S., Baek, K. H., Jung, H., Ha, S. H., Do Choi, Y., Kim, M., Reuzeau, C. and Kim, J. K. (2010). Root-specific expression of *OsNAC10* improves drought tolerance and grain yield in rice under field drought conditions. *Plant Physiol, 153*(1), 185-197. doi: 10.1104/pp.110.154773

Jiang, Y., Cai, Z., Xie, W., Long, T., Yu, H. and Zhang, Q. (2012). Rice functional genomics research: progress and implications for crop genetic improvement. *Biotechnol Adv, 30*(5), 1059-1070. doi: 10.1016/j.biotechadv.2011.08.013

Johnson, A. A., Kyriacou, B., Callahan, D. L., Carruthers, L., Stangoulis, J., Lombi, E. and Tester, M. (2011). Constitutive overexpression of the *OsNAS* gene family reveals single-gene strategies for effective iron- and zinc-biofortification of rice endosperm. *PLoS One, 6*(9), e24476. doi: 10.1371/journal.pone.0024476

Jol, S. J., Kümmel, A., Terzer, M., Stelling, J. and Heinemann, M. (2012). System-level insights into yeast metabolism by thermodynamic analysis of elementary flux modes. *PLoS Comput Biol, 8*(3), e1002415. doi: 10.1371/journal.pcbi.1002415

Joo, J., Choi, H. J., Lee, Y. H., Lee, S., Lee, C. H., Kim, C. H., Cheong, J. J., Choi, Y. D. and Song, S. I. (2014). Over-expression of *BvMTSH*, a fusion gene for maltooligosyltrehalose synthase and maltooligosyltrehalose trehalohydrolase, enhances drought tolerance in transgenic rice. *BMB Rep, 47*(1), 27-32.

Junker, A., Muraya, M. M., Weigelt-Fischer, K., Arana-Ceballos, F., Klukas, C., Melchinger, A. E., Meyer, R. C., Riewe, D. and Altmann, T. (2014). Optimizing experimental procedures for quantitative evaluation of crop plant performance in high throughput phenotyping systems. *Front Plant Sci, 5*, 770. doi: 10.3389/fpls.2014.00770

Junker, B. H. (2014). Flux analysis in plant metabolic networks: increasing throughput and coverage. *Curr Opin Biotechnol, 26*, 183-188. doi: 10.1016/j.copbio.2014.01.016

Kaldenhoff, R., Kai, L. and Uehlein, N. (2014). Aquaporins and membrane diffusion of CO_2 in living organisms. *Biochim Biophys Acta, 1840*(5), 1592-1595. doi: 10.1016/j.bbagen.2013.09.037

Kanehisa, M., Araki, M., Goto, S., Hattori, M., Hirakawa, M., Itoh, M., Katayama, T., Kawashima, S., Okuda, S., Tokimatsu, T. and Yamanishi, Y. (2008). KEGG for linking genomes to life and the environment. *Nucleic Acids Res, 36*(Database issue), D480-484. doi: 10.1093/nar/gkm882

Karmakar, R., Bhattacharya, R. and Kulshrestha, G. (2009). Comparative metabolite profiling of the insecticide thiamethoxam in plant and cell suspension culture of tomato. *J Agric Food Chem, 57*(14), 6369-6374. doi: 10.1021/jf9008394

Kauffman, K. J., Prakash, P. and Edwards, J. S. (2003). Advances in flux balance analysis. *Curr Opin Biotechnol, 14*(5), 491-496.

Kawahara, Y., de la Bastide, M., Hamilton, J. P., Kanamori, H., McCombie, W. R., Ouyang, S., Schwartz, D. C., Tanaka, T., Wu, J., Zhou, S., Childs, K. L., Davidson, R. M., Lin, H., Quesada-Ocampo, L., Vaillancourt, B., Sakai, H., Lee, S. S., Kim, J., Numa, H., Itoh, T., Buell, C. R. and Matsumoto, T. (2013). Improvement of the *Oryza sativa* Nipponbare reference genome using next generation sequence and optical map data. *Rice (N Y), 6*(1), 4. doi: 10.1186/1939-8433-6-4

KEGG. (release 65.0). from http://www.genome.jp/kegg

Kelleher, J. K. (2001). Flux estimation using isotopic tracers: common ground for metabolic physiology and metabolic engineering. *Metab Eng, 3*(2), 100-110. doi: 10.1006/mben.2001.0185

Kim, B., Park, H., Na, D. and Lee, S. Y. (2014a). Metabolic engineering of *Escherichia coli* for the production of phenol from glucose. *Biotechnol J, 9*(5), 621-629. doi: 10.1002/biot.201300263

Kim, J. G., Back, K., Lee, H. Y., Lee, H. J., Phung, T. H., Grimm, B. and Jung, S. (2014b). Increased expression of Fe-chelatase leads to increased metabolic flux into heme and confers protection against photodynamically induced oxidative stress. *Plant Mol Biol, 86*(3), 271-287. doi: 10.1007/s11103-014-0228-3

Kim, S. and Vanden Born, W. H. (1996). Chlorsulfuron decreases both assimilate export by source leaves and import by sink leaves in canola (*Brassica napus* L.) seedlings. *Pestic. Biochem. Physiol.*(56), 141-148.

Kim, S. T., Kim, S. G., Agrawal, G. K., Kikuchi, S. and Rakwal, R. (2014c). Rice proteomics: a model system for crop improvement and food security. *Proteomics, 14*(4-5), 593-610. doi: 10.1002/pmic.201300388

Kind, S., Neubauer, S., Becker, J., Yamamoto, M., Volkert, M., Abendroth, G., Zelder, O. and Wittmann, C. (2014). From zero to hero - production of bio-based nylon from renewable resources using engineered *orynebacterium glutamicum. Metab Eng, 25*, 113-123. doi: 10.1016/j.ymben.2014.05.007

Klamt, S., Gagneur, J. and von Kamp, A. (2005). Algorithmic approaches for computing elementary modes in large biochemical reaction networks. *Syst Biol (Stevenage), 152*(4), 249-255.

Klingenberg, M. (2008). The ADP and ATP transport in mitochondria and its carrier. *Biochim Biophys Acta, 1778*(10), 1978-2021. doi: 10.1016/j.bbamem.2008.04.011

Kohlstedt, M., Becker, J. and Wittmann, C. (2010). Metabolic fluxes and beyond-systems biology understanding and engineering of microbial metabolism. *Appl Microbiol Biotechnol, 88*(5), 1065-1075. doi: 10.1007/s00253-010-2854-2

Koornneef, M. and Meinke, D. (2010). The development of Arabidopsis as a model plant. *Plant J, 61*(6), 909-921. doi: 10.1111/j.1365-313X.2009.04086.x

Kramer, D. M., Avenson, T. J. and Edwards, G. E. (2004). Dynamic flexibility in the light reactions of photosynthesis governed by both electron and proton transfer reactions. *Trends Plant Sci, 9*(7), 349-357. doi: 10.1016/j.tplants.2004.05.001

Krömer, J. O., Wittmann, C., Schröder, H. and Heinzle, E. (2006). Metabolic pathway analysis for rational design of L-methionine production by *Escherichia coli* and *Corynebacterium glutamicum. Metab Eng, 8*(4), 353-369. doi: 10.1016/j.ymben.2006.02.001

Kruger, N. J., Masakapalli, S. K. and Ratcliffe, R. G. (2012). Strategies for investigating the plant metabolic network with steady-state metabolic flux analysis: lessons from an *Arabidopsis* cell culture and other systems. *J Exp Bot, 63*(6), 2309-2323. doi: 10.1093/jxb/err382

Kruger, N. J. and von Schaewen, A. (2003). The oxidative pentose phosphate pathway: structure and organisation. *Curr Opin Plant Biol, 6*(3), 236-246.

Laloi, M. (1999). Plant mitochondrial carriers: an overview. *Cell Mol Life Sci, 56*(11-12), 918-944.

Lalonde, S., Wipf, D. and Frommer, W. B. (2004). Transport mechanisms for organic forms of carbon and nitrogen between source and sink. *Annu Rev Plant Biol, 55*, 341-372. doi: 10.1146/annurev.arplant.55.031903.141758

Lassen, L. M., Nielsen, A. Z., Ziersen, B., Gnanasekaran, T., Moller, B. L. and Jensen, P. E. (2014). Redirecting photosynthetic electron flow into light-driven synthesis of alternative products including high-value bioactive natural compounds. *ACS Synth Biol, 3*(1), 1-12. doi: 10.1021/sb400136f

Lee, H. J., Abdula, S. E., Jang, D. W., Park, S. H., Yoon, U. H., Jung, Y. J., Kang, K. K., Nou, I. S. and Cho, Y. G. (2013). Overexpression of the glutamine synthetase gene modulates oxidative stress response in rice after exposure to cadmium stress. *Plant Cell Rep, 32*(10), 1521-1529. doi: 10.1007/s00299-013-1464-8

Lee, S. I., Kim, H. U., Lee, Y. H., Suh, S. C., Lim, Y. P., Lee, H. Y. and Kim, H. I. (2001). Constitutive and seed-specific expression of a maize lysine-feedback-insensitive dihydrodipicolinate synthase gene leads to increased free lysine levels in rice seeds. *Molecular Breeding, 8*(1), 75-84. doi: Doi 10.1023/A:1011977219926

Lee, T. T., Wang, M. M., Hou, R. C., Chen, L. J., Su, R. C., Wang, C. S. and Tzen, J. T. (2003). Enhanced methionine and cysteine levels in transgenic rice seeds by the accumulation of sesame 2S albumin. *Biosci Biotechnol Biochem, 67*(8), 1699-1705. doi: 10.1271/bbb.67.1699

Lehmann, M., Schwarzlander, M., Obata, T., Sirikantaramas, S., Burow, M., Olsen, C. E., Tohge, T., Fricker, M. D., Moller, B. L., Fernie, A. R., Sweetlove, L. J. and Laxa, M. (2009). The metabolic response of *Arabidopsis* roots to oxidative stress is distinct from that of heterotrophic cells in culture and highlights a complex relationship between the levels of transcripts, metabolites, and flux. *Mol Plant, 2*(3), 390-406. doi: 10.1093/mp/ssn080

Leighty, R. W. and Antoniewicz, M. R. (2012). Parallel labeling experiments with [U-13C]glucose validate *E. coli* metabolic network model for 13C metabolic flux analysis. *Metab Eng, 14*(5), 533-541. doi: 10.1016/j.ymben.2012.06.003

Li, C. R., Liang, D. D., Xu, R. F., Li, H., Zhang, Y. P., Qin, R. Y., Li, L., Wei, P. C. and Yang, J. B. (2013). Overexpression of an alternative oxidase gene, *OsAOX1a*, improves cold tolerance in *Oryza sativa* L. *Genet Mol Res, 12*(4), 5424-5432. doi: 10.4238/2013.November.11.4

Li, J. F., Zhang, D. and Sheen, J. (2015). Targeted plant genome editing via the CRISPR/Cas9 technology. *Methods Mol Biol, 1284*, 239-255. doi: 10.1007/978-1-4939-2444-8_12

Liang, X., Zhang, L., Natarajan, S. K. and Becker, D. F. (2013). Proline mechanisms of stress survival. *Antioxid Redox Signal, 19*(9), 998-1011. doi: 10.1089/ars.2012.5074

Libourel, I. G. and Shachar-Hill, Y. (2008). Metabolic flux analysis in plants: from intelligent design to rational engineering. *Annu Rev Plant Biol, 59*, 625-650. doi: 10.1146/annurev.arplant.58.032806.103822

Lin, M. T., Occhialini, A., Andralojc, P. J., Parry, M. A. and Hanson, M. R. (2014). A faster Rubisco with potential to increase photosynthesis in crops. *Nature, 513*(7519), 547-550. doi: 10.1038/nature13776

Liu, L., Mei, Q., Yu, Z., Sun, T., Zhang, Z. and Chen, M. (2013). An integrative bioinformatics framework for genome-scale multiple level network reconstruction of rice. *J Integr Bioinform, 10*(2), 223. doi: 10.2390/biecoll-jib-2013-223

Long, S. P. (2014). We need winners in the race to increase photosynthesis in rice, whether from conventional breeding, biotechnology or both. *Plant Cell and Environment, 37*(1), 19-21. doi: Doi 10.1111/Pce.12193

Long, S. P., Postl, W. F. and Bolharnordenkampf, H. R. (1993). Quantum Yields for Uptake of Carbon-Dioxide in C-3 Vascular Plants of Contrasting Habitats and Taxonomic Groupings. *Planta, 189*(2), 226-234.

Long, X., Liu, Q., Chan, M., Wang, Q. and Sun, S. S. (2013). Metabolic engineering and profiling of rice with increased lysine. *Plant Biotechnol J, 11*(4), 490-501. doi: 10.1111/pbi.12037

Lotz, K., Hartmann, A., Grafahrend-Belau, E., Schreiber, F. and Junker, B. H. (2014). Elementary flux modes, flux balance analysis, and their application to plant metabolism. *Methods Mol Biol, 1083*, 231-252. doi: 10.1007/978-1-62703-661-0_14

Lu, F., Wang, H., Wang, S., Jiang, W., Shan, C., Li, B., Yang, J., Zhang, S. and Sun, W. (2014). Enhancement of innate immune system in monocot rice by transferring the dicotyledonous elongation factor Tu receptor EFR. *J Integr Plant Biol.* doi: 10.1111/jipb.12306

Ma, F., Jazmin, L. J., Young, J. D. and Allen, D. K. (2014). Isotopically nonstationary ^{13}C flux analysis of changes in *Arabidopsis thaliana* leaf metabolism due to high light acclimation. *Proc Natl Acad Sci U S A, 111*(47), 16967-16972. doi: 10.1073/pnas.1319485111

Ma, J., Song, Y., Wu, B., Jiang, M., Li, K., Zhu, C. and Wen, F. (2011). Production of transgenic rice new germplasm with strong resistance against two isolations of *Rice stripe virus* by RNA interference. *Transgenic Res, 20*(6), 1367-1377. doi: 10.1007/s11248-011-9502-1

MacDonald, M. J. (1995). Feasibility of a mitochondrial pyruvate malate shuttle in pancreatic islets. Further implication of cytosolic NADPH in insulin secretion. *J Biol Chem, 270*(34), 20051-20058.

Madsen, K., Nielsen, H. B. and Tingleff, O. (2004). *Methods for non-linear least squares problems* (2nd ed.): Technical University of Denmark.

Manabe, Y., Tinker, N., Colville, A. and Miki, B. (2007). CSR1, the sole target of imidazolinone herbicide in *Arabidopsis thaliana*. *Plant Cell Physiol, 48*(9), 1340-1358. doi: 10.1093/pcp/pcm105

Masakapalli, S. K., Bryant, F. M., Kruger, N. J. and Ratcliffe, R. G. (2014). The metabolic flux phenotype of heterotrophic *Arabidopsis* cells reveals a flexible balance between the cytosolic and plastidic contributions to carbohydrate oxidation in response to phosphate limitation. *Plant J, 78*(6), 964-977. doi: 10.1111/tpj.12522

Masuda, H., Ishimaru, Y., Aung, M. S., Kobayashi, T., Kakei, Y., Takahashi, M., Higuchi, K., Nakanishi, H. and Nishizawa, N. K. (2012). Iron biofortification in rice by the introduction of multiple genes involved in iron nutrition. *Sci Rep, 2*, 543. doi: 10.1038/srep00543

McFarlane, H., Döring, A. and Persson, S. (2014). The Cell Biology of Cellulose Synthesis. *Annu Rev Plant Biol, 65*, 69-94.

Melzer, G., Esfandabadi, M. E., Franco-Lara, E. and Wittmann, C. (2009). Flux Design: In silico design of cell factories based on correlation of pathway fluxes to desired properties. *BMC Syst Biol, 3*, 120. doi: 10.1186/1752-0509-3-120

Merchant, S. S., Prochnik, S. E., Vallon, O., Harris, E. H., Karpowicz, S. J., Witman, G. B., Terry, A., Salamov, A., Fritz-Laylin, L. K., Maréchal-Drouard, L., Marshall, W. F., Qu, L. H., Nelson, D. R., Sanderfoot, A. A., Spalding, M. H., Kapitonov, V. V., Ren, Q., Ferris, P., Lindquist, E., Shapiro, H., Lucas, S. M., Grimwood, J., Schmutz, J., Cardol, P., Cerutti, H., Chanfreau, G., Chen, C. L., Cognat, V., Croft, M. T., Dent, R., Dutcher, S., Fernández, E., Fukuzawa, H., González-Ballester, D., González-Halphen, D., Hallmann, A., Hanikenne, M., Hippler, M., Inwood, W., Jabbari, K., Kalanon, M., Kuras, R., Lefebvre, P. A., Lemaire, S. D., Lobanov, A. V., Lohr, M., Manuell, A., Meier, I., Mets, L., Mittag, M., Mittelmeier, T., Moroney, J. V., Moseley, J., Napoli, C., Nedelcu, A.

M., Niyogi, K., Novoselov, S. V., Paulsen, I. T., Pazour, G., Purton, S., Ral, J. P., Riaño-Pachón, D. M., Riekhof, W., Rymarquis, L., Schroda, M., Stern, D., Umen, J., Willows, R., Wilson, N., Zimmer, S. L., Allmer, J., Balk, J., Bisova, K., Chen, C. J., Elias, M., Gendler, K., Hauser, C., Lamb, M. R., Ledford, H., Long, J. C., Minagawa, J., Page, M. D., Pan, J., Pootakham, W., Roje, S., Rose, A., Stahlberg, E., Terauchi, A. M., Yang, P., Ball, S., Bowler, C., Dieckmann, C. L., Gladyshev, V. N., Green, P., Jorgensen, R., Mayfield, S., Mueller-Roeber, B., Rajamani, S., Sayre, R. T., Brokstein, P., Dubchak, I., Goodstein, D., Hornick, L., Huang, Y. W., Jhaveri, J., Luo, Y., Martinez, D., Ngau, W. C., Otillar, B., Poliakov, A., Porter, A., Szajkowski, L., Werner, G., Zhou, K., Grigoriev, I. V., Rokhsar, D. S. and Grossman, A. R. (2007). The *Chlamydomonas* genome reveals the evolution of key animal and plant functions. *Science, 318*(5848), 245-250. doi: 10.1126/science.1143609

MetaCrop. (release 2.0, 2014). from hp://metacrop.ipk.gatersleben.de

Michael, T. P. and Jackson, S. (2013). The First 50 Plant Genomes. *Plant Genome, 6*(2). doi: DOI 10.3835/plantgenome2013.03.0001in

Mintz-Oron, S., Meir, S., Malitsky, S., Ruppin, E., Aharoni, A. and Shlomi, T. (2012). Reconstruction of *Arabidopsis* metabolic network models accounting for subcellular compartmentalization and tissue-specificity. *Proc Natl Acad Sci U S A, 109*(1), 339-344. doi: 10.1073/pnas.1100358109

Misra, S. C., Randive, R., Rao, V. S., Sheshshayee, M. S., Serraj, R. and Monneveux, P. (2006). Relationship between carbon isotope discrimination, ash content and grain yield in wheat in the Peninsular Zone of India. *Journal of Agronomy and Crop Science, 192*(5), 352-362. doi: DOI 10.1111/j.1439-037X.2006.00225.x

Moisset, P., Vaisman, D., Cintolesi, A., Urrutia, J., Rapaport, I., Andrews, B. A. and Asenjo, J. A. (2012). Continuous modeling of metabolic networks with gene regulation in yeast and *in vivo* determination of rate parameters. *Biotechnol Bioeng, 109*(9), 2325-2339. doi: 10.1002/bit.24503

Molina, J., Sikora, M., Garud, N., Flowers, J. M., Rubinstein, S., Reynolds, A., Huang, P., Jackson, S., Schaal, B. A., Bustamante, C. D., Boyko, A. R. and Purugganan, M. D. (2011). Molecular evidence for a single evolutionary origin of domesticated rice. *Proc Natl Acad Sci U S A, 108*(20), 8351-8356. doi: 10.1073/pnas.1104686108

Mora-Garcia, S., Stolowicz, F. G. and Wolosiuk, R. A. (2006). *Redox signal transduction in plant metabolism* (Vol. 22: Control of Primary Metabolism in Plants): Blackwell Publishing Ltd, Oxford, UK.

Mueller, L. A., Zhang, P. and Rhee, S. Y. (2003). AraCyc: a biochemical pathway database for Arabidopsis. *Plant Physiol, 132*(2), 453-460. doi: 10.1104/pp.102.017236

Munekage, Y. N., Genty, B. and Peltier, G. (2008). Effect of PGR5 impairment on photosynthesis and growth in *Arabidopsis thaliana*. *Plant Cell Physiol, 49*(11), 1688-1698. doi: 10.1093/pcp/pcn140

Murray, M. G. and Thompson, W. F. (1980). Rapid isolation of high molecular weight plant DNA. *Nucleic Acids Res, 8*(19), 4321-4325.

Ngounou Wetie, A. G., Sokolowska, I., Woods, A. G., Wormwood, K. L., Dao, S., Patel, S., Clarkson, B. D. and Darie, C. C. (2013). Automated mass spectrometry-based functional assay for the routine analysis of the secretome. *J Lab Autom, 18*(1), 19-29. doi: 10.1177/2211068212454738

Noctor, G. and Foyer, C. H. (2000). Homeostasis of adenylate status during photosynthesis in a fluctuating environment. *J Exp Bot, 51 Spec No*, 347-356.

Nouchi, I., Ito, O., Harazono, Y. and Kouchi, H. (1995). Acceleration of ^{13}C-labelled photosynthate partitioning from leaves to panicles in rice plants exposed to chronic ozone at the reproductive stage. *Environ Pollut, 88*(3), 253-260.

Nowicka, B., Strzalka, W. and Strzalka, K. (2009). New transgenic line of *Arabidopsis thaliana* with partly disabled zeaxanthin epoxidase activity displays changed carotenoid composition, xanthophyll cycle activity and non-photochemical quenching kinetics. *J Plant Physiol, 166*(10), 1045-1056. doi: 10.1016/j.jplph.2008.12.010

Núñez, O., Gallart-Ayala, H., Martins, C. P., Lucci, P. and Busquets, R. (2013). State-of-the-art in fast liquid chromatography-mass spectrometry for bio-analytical applications. *J Chromatogr B Analyt Technol Biomed Life Sci, 927*, 3-21. doi: 10.1016/j.jchromb.2012.12.031

OryzaCyc. (version 2.0, 2014). from http:pmn.plantcyc.org

Orzechowski, S. (2008). Starch metabolism in leaves. *Acta Biochim Pol, 55*(3), 435-445.

Osborne, B. A. and Garrett, M. K. (1983). Quantum yield for CO_2 uptake in some diploid and tetraploid plant species. *Plant, Cell and Environment, 6*, 135-144.

Paddon, C. J., Westfall, P. J., Pitera, D. J., Benjamin, K., Fisher, K., McPhee, D., Leavell, M. D., Tai, A., Main, A., Eng, D., Polichuk, D. R., Teoh, K. H., Reed, D. W., Treynor, T., Lenihan, J., Fleck, M., Bajad, S., Dang, G., Dengrove, D., Diola, D., Dorin, G., Ellens, K. W., Fickes, S., Galazzo, J., Gaucher, S. P., Geistlinger, T., Henry, R., Hepp, M., Horning, T., Iqbal, T., Jiang, H., Kizer, L., Lieu, B., Melis, D., Moss, N., Regentin, R., Secrest, S., Tsuruta, H., Vazquez, R., Westblade, L. F., Xu, L., Yu, M., Zhang, Y., Zhao, L., Lievense, J., Covello, P. S., Keasling, J. D., Reiling, K. K., Renninger, N. S. and Newman, J. D. (2013). High-level semi-synthetic production of the potent antimalarial artemisinin. *Nature, 496*(7446), 528-532. doi: 10.1038/nature12051

Panda, D. and Sarkar, R. K. (2013). Natural leaf senescence: probed by chlorophyll fluorescence, CO_2 photosynthetic rate and antioxidant enzyme activities during grain filling in different rice cultivars. *Physiol Mol Biol Plants, 19*(1), 43-51. doi: 10.1007/s12298-012-0142-6

Papin, J. A., Price, N. D. and Palsson, B. Ø. (2002). Extreme pathway lengths and reaction participation in genome-scale metabolic networks. *Genome Res, 12*(12), 1889-1900. doi: 10.1101/gr.327702

Park, S., Lee, D. E., Jang, H., Byeon, Y., Kim, Y. S. and Back, K. (2013). Melatonin-rich transgenic rice plants exhibit resistance to herbicide-induced oxidative stress. *J Pineal Res, 54*(3), 258-263. doi: 10.1111/j.1600-079X.2012.01029.x

Parry, M. A., Andralojc, P. J., Scales, J. C., Salvucci, M. E., Carmo-Silva, A. E., Alonso, H. and Whitney, S. M. (2013). Rubisco activity and regulation as targets for crop improvement. *J Exp Bot, 64*(3), 717-730. doi: 10.1093/jxb/ers336

Paterson, A. H., Bowers, J. E., Bruggmann, R., Dubchak, I., Grimwood, J., Gundlach, H., Haberer, G., Hellsten, U., Mitros, T., Poliakov, A., Schmutz, J., Spannagl, M., Tang, H., Wang, X., Wicker, T., Bharti, A. K., Chapman, J., Feltus, F. A., Gowik, U., Grigoriev, I. V., Lyons, E., Maher, C. A., Martis, M., Narechania, A., Otillar, R. P., Penning, B. W., Salamov, A. A., Wang, Y., Zhang, L., Carpita, N. C., Freeling, M., Gingle, A. R., Hash, C. T., Keller, B., Klein, P., Kresovich, S., McCann, M. C., Ming, R., Peterson, D. G., Mehboob ur, R., Ware, D., Westhoff, P., Mayer, K. F., Messing, J.

and Rokhsar, D. S. (2009). The *Sorghum bicolor* genome and the diversification of grasses. *Nature, 457*(7229), 551-556. doi: 10.1038/nature07723

Pauly, M., Gille, S., Liu, L., Mansoori, N., de Souza, A., Schultink, A. and Xiong, G. (2013). Hemicellulose biosynthesis. *Planta, 238*(4), 627-642. doi: 10.1007/s00425-013-1921-1

Petolino, J. F. (2015). Genome editing in plants via designed zinc finger nucleases. *In Vitro Cell Dev Biol Plant, 51*(1), 1-8. doi: 10.1007/s11627-015-9663-3

Pfeiffer, T., Sánchez-Valdenebro, I., Nuño, J. C., Montero, F. and Schuster, S. (1999). METATOOL: for studying metabolic networks. *Bioinformatics, 15*(3), 251-257.

Picault, N., Hodges, M., Palmieri, L. and Palmieri, F. (2004). The growing family of mitochondrial carriers in *Arabidopsis*. *Trends Plant Sci, 9*(3), 138-146. doi: 10.1016/j.tplants.2004.01.007

Pilalis, E., Chatziioannou, A., Thomasset, B. and Kolisis, F. (2011). An *in silico* compartmentalized metabolic model of *Brassica napus* enables the systemic study of regulatory aspects of plant central metabolism. *Biotechnol Bioeng, 108*(7), 1673-1682. doi: 10.1002/bit.23107

Plaxton, W. C. (1996). The Organization and Regulation of Plant Glycolysis. *Annu Rev Plant Physiol Plant Mol Biol, 47*, 185-214. doi: 10.1146/annurev.arplant.47.1.185

Plaxton, W. C. and Tran, H. T. (2011). Metabolic adaptations of phosphate-starved plants. *Plant Physiol, 156*(3), 1006-1015. doi: 10.1104/pp.111.175281

Poblete-Castro, I., Binger, D., Rodrigues, A., Becker, J., Martins Dos Santos, V. A. and Wittmann, C. (2013). In-silico-driven metabolic engineering of *Pseudomonas putida* for enhanced production of poly-hydroxyalkanoates. *Metab Eng, 15*, 113-123. doi: 10.1016/j.ymben.2012.10.004

Poolman, M. G., Kundu, S., Shaw, R. and Fell, D. A. (2013). Responses to light intensity in a genome-scale model of rice metabolism. *Plant Physiol, 162*(2), 1060-1072. doi: 10.1104/pp.113.216762

Poolman, M. G., Kundu, S., Shaw, R. and Fell, D. A. (2014). Metabolic trade-offs between biomass synthesis and photosynthate export at different light intensities in a genome-scale metabolic model of rice. *Front Plant Sci, 5*, 656. doi: 10.3389/fpls.2014.00656

Poolman, M. G., Miguet, L., Sweetlove, L. J. and Fell, D. A. (2009). A genome-scale metabolic model of Arabidopsis and some of its properties. *Plant Physiol, 151*(3), 1570-1581. doi: 10.1104/pp.109.141267

Poschet, G., Hannich, B., Raab, S., Jungkunz, I., Klemens, P. A., Krueger, S., Wic, S., Neuhaus, H. E. and Büttner, M. (2011). A novel Arabidopsis vacuolar glucose exporter is involved in cellular sugar homeostasis and affects the composition of seed storage compounds. *Plant Physiol, 157*(4), 1664-1676. doi: 10.1104/pp.111.186825

Price, G. D., Pengelly, J. J., Forster, B., Du, J., Whitney, S. M., von Caemmerer, S., Badger, M. R., Howitt, S. M. and Evans, J. R. (2013). The cyanobacterial CCM as a source of genes for improving photosynthetic CO_2 fixation in crop species. *J Exp Bot, 64*(3), 753-768. doi: 10.1093/jxb/ers257

Price, N. D., Papin, J. A. and Palsson, B. Ø. (2002). Determination of redundancy and systems properties of the metabolic network of *Helicobacter pylori* using genome-scale extreme pathway analysis. *Genome Res, 12*(5), 760-769. doi: 10.1101/gr.218002. Article published online before print in April 2002

Qi, Y. B., Ye, S. H., Lu, Y. T., Jin, Q. S. and Zhang, X. M. (2009). Development of marker-free transgenic *Cry1Ab* Rice with Lepidopteran pest resistance by *Agrobacterium* mixture-mediated co-transformation. *Rice Science, 16*(3), 181-186.

Qian, B., Li, X., Liu, X., Chen, P., Ren, C. and Dai, C. (2015). Enhanced drought tolerance in transgenic rice over-expressing of maize C_4 phosphoenolpyruvate carboxylase gene via NO and Ca^{2+}. *J Plant Physiol, 175*, 9-20. doi: 10.1016/j.jplph.2014.09.019

Qian, Q., Huang, L., Yi, R., Wang, S. and Ding, Y. (2014). Enhanced resistance to blast fungus in rice (*Oryza sativa* L.) by expressing the ribosome-inactivating protein alpha-momorcharin. *Plant Sci, 217-218*, 1-7. doi: 10.1016/j.plantsci.2013.11.012

Qin, D., Wang, F., Geng, X., Zhang, L., Yao, Y., Ni, Z., Peng, H. and Sun, Q. (2015). Overexpression of heat stress-responsive *TaMBF1c*, a wheat (*Triticum aestivum* L.) Multiprotein Bridging Factor, confers heat tolerance in both yeast and rice. *Plant Mol Biol, 87*(1-2), 31-45. doi: 10.1007/s11103-014-0259-9

Quek, L. E., Wittmann, C., Nielsen, L. K. and Krömer, J. O. (2009). OpenFLUX: efficient modelling software for ^{13}C-based metabolic flux analysis. *Microb Cell Fact, 8*, 25. doi: 10.1186/1475-2859-8-25

Quilis, J., López-García, B., Meynard, D., Guiderdoni, E. and San Segundo, B. (2014). Inducible expression of a fusion gene encoding two proteinase inhibitors leads to insect and pathogen resistance in transgenic rice. *Plant Biotechnol J, 12*(3), 367-377. doi: 10.1111/pbi.12143

Raghavendra, A. S., Reumann, S. and Heldt, H. W. (1998). Participation of mitochondrial metabolism in photorespiration. Reconstituted system of peroxisomes and mitochondria from spinach leaves. *Plant Physiol, 116*(4), 1333-1337.

Raghothama, K. G. (1999). Phosphate Acquisition. *Annu Rev Plant Physiol Plant Mol Biol, 50*, 665-693. doi: 10.1146/annurev.arplant.50.1.665

Rajasekaran, A. and Kalaivani, M. (2013). Designer foods and their benefits: A review. *J Food Sci Technol, 50*(1), 1-16. doi: 10.1007/s13197-012-0726-8

Ratcliffe, R. G. and Shachar-Hill, Y. (2006). Measuring multiple fluxes through plant metabolic networks. *Plant J, 45*(4), 490-511. doi: 10.1111/j.1365-313X.2005.02649.x

Rausch, C. and Bucher, M. (2002). Molecular mechanisms of phosphate transport in plants. *Planta, 216*(1), 23-37. doi: 10.1007/s00425-002-0921-3

Rawsthorne, S. (2002). Carbon flux and fatty acid synthesis in plants. *Prog Lipid Res, 41*(2), 182-196.

Ray, D. K., Mueller, N. D., West, P. C. and Foley, J. A. (2013). Yield Trends Are Insufficient to Double Global Crop Production by 2050. *PLoS One, 8*(6), e66428. doi: 10.1371/journal.pone.0066428

Reiter, W. D. (2002). Biosynthesis and properties of the plant cell wall. *Curr Opin Plant Biol, 5*(6), 536-542.

Reiter, W. D., Chapple, C. and Somerville, C. R. (1997). Mutants of *Arabidopsis thaliana* with altered cell wall polysaccharide composition. *Plant J, 12*(2), 335-345.

Reyes-Prieto, A. and Moustafa, A. (2012). Plastid-localized amino acid biosynthetic pathways of Plantae are predominantly composed of non-cyanobacterial enzymes. *Sci Rep, 2*, 955. doi: 10.1038/srep00955

Rezola, A., de Figueiredo, L. F., Brock, M., Pey, J., Podhorski, A., Wittmann, C., Schuster, S., Bockmayr, A. and Planes, F. J. (2011). Exploring metabolic pathways in genome-scale networks via generating flux modes. *Bioinformatics, 27*(4), 534-540. doi: 10.1093/bioinformatics/btq681

Rodríguez-Concepción, M. and Boronat, A. (2002). Elucidation of the methylerythritol phosphate pathway for isoprenoid biosynthesis in bacteria and plastids. A metabolic milestone achieved through genomics. *Plant Physiol, 130*(3), 1079-1089. doi: 10.1104/pp.007138

Roessner, U., Wagner, C., Kopka, J., Trethewey, R. N. and Willmitzer, L. (2000). Technical advance: simultaneous analysis of metabolites in potato tuber by gas chromatography-mass spectrometry. *Plant J, 23*(1), 131-142.

Rohwer, J. M. (2012). Kinetic modelling of plant metabolic pathways. *J Exp Bot, 63*(6), 2275-2292. doi: 10.1093/jxb/ers080

Rohwer, J. M. and Botha, F. C. (2001). Analysis of sucrose accumulation in the sugar cane culm on the basis of *in vitro* kinetic data. *Biochem J, 358*(Pt 2), 437-445.

Römisch-Margl, W., Schramek, N., Radykewicz, T., Ettenhuber, C., Eylert, E., Huber, C., Römisch-Margl, L., Schwarz, C., Dobner, M., Demmel, N., Winzenhörlein, B., Bacher, A. and Eisenreich, W. (2007). $^{13}CO_2$ as a universal metabolic tracer in isotopologue perturbation experiments. *Phytochemistry, 68*(16-18), 2273-2289. doi: 10.1016/j.phytochem.2007.03.034

Roosens, N. H., Thu, T. T., Iskandar, H. M. and Jacobs, M. (1998). Isolation of the ornithine-delta-aminotransferase cDNA and effect of salt stress on its expression in *Arabidopsis thaliana*. *Plant Physiol, 117*(1), 263-271.

Royuela, M., Gonzalez, A., Gonzalez, E. M., Arrese-Igor, C., Aparicio-Tejo, P. M. and Gonzalez-Murua, C. (2000). Physiological consequences of continuous, sublethal imazethapyr supply to pea plants. *J Plant Physiol, 157*(3), 345-354.

Saha, R., Suthers, P. F. and Maranas, C. D. (2011). *Zea mays* iRS1563: a comprehensive genome-scale metabolic reconstruction of maize metabolism. *PLoS One, 6*(7), e21784. doi: 10.1371/journal.pone.0021784

Saha, S. and Ramachandran, S. (2013). Genetic improvement of plants for enhanced bio-ethanol production. *Recent Pat DNA Gene Seq, 7*(1), 36-44.

Sahoo, R. K., Ansari, M. W., Tuteja, R. and Tuteja, N. (2014). *OsSUV3* transgenic rice maintains higher endogenous levels of plant hormones that mitigates adverse effects of salinity and sustains crop productivity. *Rice (N Y), 7*(1), 17. doi: 10.1186/s12284-014-0017-2

Sakamoto, T. and Matsuoka, M. (2004). Generating high-yielding varieties by genetic manipulation of plant architecture. *Curr Opin Biotechnol, 15*(2), 144-147. doi: 10.1016/j.copbio.2004.02.003

Sang, T. and Ge, S. (2007). Genetics and phylogenetics of rice domestication. *Curr Opin Genet Dev,* *17*(6), 533-538. doi: 10.1016/j.gde.2007.09.005

Sasaya, T., Nakazono-Nagaoka, E., Saika, H., Aoki, H., Hiraguri, A., Netsu, O., Uehara-Ichiki, T., Onuki, M., Toki, S., Saito, K. and Yatou, O. (2014). Transgenic strategies to confer resistance against viruses in rice plants. *Front Microbiol, 4*, 409. doi: 10.3389/fmicb.2013.00409

Sauer, U. (2006). Metabolic networks in motion: ^{13}C-based flux analysis. *Mol Syst Biol, 2*, 62. doi: 10.1038/msb4100109

Saxena, S. C., Salvi, P., Kaur, H., Verma, P., Petla, B. P., Rao, V., Kamble, N. and Majee, M. (2013). Differentially expressed *myo*-inositol monophosphatase gene (*CaIMP*) in chickpea (*Cicer arietinum* L.) encodes a lithium-sensitive phosphatase enzyme with broad substrate specificity and improves seed germination and seedling growth under abiotic stresses. *J Exp Bot, 64*(18), 5623-5639. doi: 10.1093/jxb/ert336

Schaefer, J., Kier, L. D. and Stejskal, E. O. (1980). Characterization of photorespiration in intact leaves using carbon dioxide labeling. *Plant Physiol, 65*(2), 254-259.

Schallau, K. and Junker, B. H. (2010). Simulating plant metabolic pathways with enzyme-kinetic models. *Plant Physiol, 152*(4), 1763-1771. doi: 10.1104/pp.109.149237

Schäuble, S., Heiland, I., Voytsekh, O., Mittag, M. and Schuster, S. (2011). Predicting the physiological role of circadian metabolic regulation in the green alga *Chlamydomonas reinhardtii*. *PLoS One, 6*(8), e23026. doi: 10.1371/journal.pone.0023026

Schnarrenberger, C. and Martin, W. (2002). Evolution of the enzymes of the citric acid cycle and the glyoxylate cycle of higher plants. A case study of endosymbiotic gene transfer. *Eur J Biochem, 269*(3), 868-883.

Schuster, S., Dandekar, T. and Fell, D. A. (1999). Detection of elementary flux modes in biochemical networks: a promising tool for pathway analysis and metabolic engineering. *Trends Biotechnol, 17*(2), 53-60.

Schwender, J., Goffman, F., Ohlrogge, J. B. and Shachar-Hill, Y. (2004). Rubisco without the Calvin cycle improves the carbon efficiency of developing green seeds. *Nature, 432*(7018), 779-782. doi: 10.1038/nature03145

Schwender, J., Shachar-Hill, Y. and Ohlrogge, J. B. (2006). Mitochondrial metabolism in developing embryos of *Brassica napus*. *J Biol Chem, 281*(45), 34040-34047. doi: 10.1074/jbc.M606266200

Seaver, S. M., Bradbury, L. M., Frelin, O., Zarecki, R., Ruppin, E., Hanson, A. D. and Henry, C. S. (2015). Improved evidence-based genome-scale metabolic models for maize leaf, embryo, and endosperm. *Front Plant Sci, 6*, 142. doi: 10.3389/fpls.2015.00142

Shachar-Hill, Y. (2013). Metabolic network flux analysis for engineering plant systems. *Curr Opin Biotechnol, 24*(2), 247-255. doi: 10.1016/j.copbio.2013.01.004

Shaner, D. L. and Singh, B. K. (1991). Imidazolinone-induced loss of acetohydroxyacid synthase activity in maize is not due to the enzyme degradation. *Plant Physiol, 97*(4), 1339-1341.

Shen, W., Li, J. Q., Dauk, M., Huang, Y., Periappuram, C., Wei, Y. and Zou, J. (2010). Metabolic and transcriptional responses of glycerolipid pathways to a perturbation of glycerol 3-phosphate metabolism in *Arabidopsis. J Biol Chem, 285*(30), 22957-22965. doi: 10.1074/jbc.M109.097758

Shimizu, T., Ogamino, T., Hiraguri, A., Nakazono-Nagaoka, E., Uehara-Ichiki, T., Nakajima, M., Akutsu, K., Omura, T. and Sasaya, T. (2013). Strong resistance against *Rice grassy stunt virus* is induced in transgenic rice plants expressing double-stranded RNA of the viral genes for nucleocapsid or movement proteins as targets for RNA interference. *Phytopathology, 103*(5), 513-519. doi: 10.1094/PHYTO-07-12-0165-R

Shubhakar, A., Reiding, K. R., Gardner, R. A., Spencer, D. I., Fernandes, D. L. and Wuhrer, M. (2015). High-Throughput Analysis and Automation for Glycomics Studies. *Chromatographia, 78*(5-6), 321-333. doi: 10.1007/s10337-014-2803-9

Simons, M., Saha, R., Amiour, N., Kumar, A., Guillard, L., Clément, G., Miquel, M., Li, Z., Mouille, G., Lea, P. J., Hirel, B. and Maranas, C. D. (2014). Assessing the metabolic impact of nitrogen availability using a compartmentalized maize leaf genome-scale model. *Plant Physiol, 166*(3), 1659-1674. doi: 10.1104/pp.114.245787

Skillman, J. B. (2008). Quantum yield variation across the three pathways of photosynthesis: not yet out of the dark. *J Exp Bot, 59*(7), 1647-1661. doi: Doi 10.1093/Jxb/Ern029

Smidansky, E. D., Martin, J. M., Hannah, L. C., Fischer, A. M. and Giroux, M. J. (2003). Seed yield and plant biomass increases in rice are conferred by deregulation of endosperm ADP-glucose pyrophosphorylase. *Planta, 216*(4), 656-664. doi: 10.1007/s00425-002-0897-z

Smith, A. M. and Stitt, M. (2007). Coordination of carbon supply and plant growth. *Plant Cell Environ, 30*(9), 1126-1149. doi: 10.1111/j.1365-3040.2007.01708.x

Sokol, S., Millard, P. and Portais, J. C. (2012). influx_s: increasing numerical stability and precision for metabolic flux analysis in isotope labelling experiments. *Bioinformatics, 28*(5), 687-693. doi: 10.1093/bioinformatics/btr716

Sørhagen, K., Laxa, M., Peterhänsel, C. and Reumann, S. (2013). The emerging role of photorespiration and non-photorespiratory peroxisomal metabolism in pathogen defence. *Plant Biol (Stuttg), 15*(4), 723-736. doi: 10.1111/j.1438-8677.2012.00723.x

Srour, O., Young, J. D. and Eldar, Y. C. (2011). Fluxomers: a new approach for ^{13}C metabolic flux analysis. *BMC Syst Biol, 5*, 129. doi: 10.1186/1752-0509-5-129

Stahl, U., Carlsson, A. S., Lenman, M., Dahlqvist, A., Huang, B., Banas, W., Banas, A. and Stymne, S. (2004). Cloning and functional characterization of a phospholipid:diacylglycerol acyltransferase from *Arabidopsis. Plant Physiol, 135*(3), 1324-1335. doi: 10.1104/pp.104.044354

Steer, B. T. (1974). Control of diurnal variations in photosynthetic products: I. Carbon metabolism. *Plant Physiol, 54*(5), 758-761.

Stephanopoulos, G. (1999). Metabolic fluxes and metabolic engineering. *Metab Eng, 1*(1), 1-11. doi: 10.1006/mben.1998.0101

Stitt, M., Lilley, R., Gerhardt, R. and Heldt, H. (1989). Determination of metabolite levels in specific cells and subcellular compartments of plant leaves. *Methods Enzymology, 174*, 518-552.

Storozhenko, S., De Brouwer, V., Volckaert, M., Navarrete, O., Blancquaert, D., Zhang, G. F., Lambert, W. and Van Der Straeten, D. (2007). Folate fortification of rice by metabolic engineering. *Nat Biotechnol, 25*(11), 1277-1279. doi: 10.1038/nbt1351

Streb, S. and Zeeman, S. C. (2012). Starch metabolism in *Arabidopsis*. *Arabidopsis Book, 10*, e0160. doi: 10.1199/tab.0160

Sulpice, R., Flis, A., Ivakov, A. A., Apelt, F., Krohn, N., Encke, B., Abel, C., Feil, R., Lunn, J. E. and Stitt, M. (2014). *Arabidopsis* coordinates the diurnal regulation of carbon allocation and growth across a wide range of photoperiods. *Mol Plant, 7*(1), 137-155. doi: 10.1093/mp/sst127

Sumiyoshi, M., Nakamura, A., Nakamura, H., Hakata, M., Ichikawa, H., Hirochika, H., Ishii, T., Satoh, S. and Iwai, H. (2013). Increase in cellulose accumulation and improvement of saccharification by overexpression of arabinofuranosidase in rice. *PLoS One, 8*(11), e78269. doi: 10.1371/journal.pone.0078269

Suthers, P. F., Chang, Y. J. and Maranas, C. D. (2010). Improved computational performance of MFA using elementary metabolite units and flux coupling. *Metab Eng, 12*(2), 123-128. doi: 10.1016/j.ymben.2009.10.002

Suzuki, K., Kaminuma, O., Yang, L., Takai, T., Mori, A., Umezu-Goto, M., Ohtomo, T., Ohmachi, Y., Noda, Y., Hirose, S., Okumura, K., Ogawa, H., Takada, K., Hirasawa, M., Hiroi, T. and Takaiwa, F. (2011). Prevention of allergic asthma by vaccination with transgenic rice seed expressing mite allergen: induction of allergen-specific oral tolerance without bystander suppression. *Plant Biotechnol J, 9*(9), 982-990. doi: 10.1111/j.1467-7652.2011.00613.x

Suzuki, Y., Kawazu, T. and Koyama, H. (2004). RNA isolation from siliques, dry seeds, and other tissues of *Arabidopsis thaliana*. *Biotechniques, 37*(4), 542, 544.

Suzuki, Y., Makino, A. and Mae, T. (2001). An efficient method for extraction of RNA from rice leaves at different ages using benzyl chloride. *J Exp Bot, 52*(360), 1575-1579.

Sweetlove, L. J. and Fernie, A. R. (2013). The spatial organization of metabolism within the plant cell. *Annu Rev Plant Biol, 64*, 723-746. doi: 10.1146/annurev-arplant-050312-120233

Sweetlove, L. J., Last, R. L. and Fernie, A. R. (2003). Predictive metabolic engineering: a goal for systems biology. *Plant Physiol, 132*(2), 420-425. doi: 10.1104/pp.103.022004

Szecowka, M., Heise, R., Tohge, T., Nunes-Nesi, A., Vosloh, D., Huege, J., Feil, R., Lunn, J., Nikoloski, Z., Stitt, M., Fernie, A. R. and Arrivault, S. (2013). Metabolic fluxes in an illuminated *Arabidopsis* rosette. *Plant Cell, 25*(2), 694-714. doi: 10.1105/tpc.112.106989

Taffs, R., Aston, J. E., Brileya, K., Jay, Z., Klatt, C. G., McGlynn, S., Mallette, N., Montross, S., Gerlach, R., Inskeep, W. P., Ward, D. M. and Carlson, R. P. (2009). *In silico* approaches to study mass and energy flows in microbial consortia: a syntrophic case study. *BMC Syst Biol, 3*, 114. doi: 10.1186/1752-0509-3-114

TAIR. The Arabidopsis Information Resource. Retrieved 27.04.2015, from http://www.arabidopsis.org

Taiz, L. and Zeiger, E. (2006). *Plant Physiology* (4th ed.). Sunderland, Massachusetts: Sinauer Associates, Inc.

Takahashi, M. (2003). Overcoming Fe deficiency by a transgenic approach in rice. *Plant Cell Tissue and Organ Culture, 72*(3), 211-220. doi: Doi 10.1023/A:1022304106367

Tanaka, A. and Osaki, M. (1983). Growth and Behavior of Photosynthesized C-14 in Various Crops in Relation to Productivity. *Soil Science and Plant Nutrition, 29*(2), 147-158.

Tardy, F. and Havaux, M. (1996). Photosynthesis, chlorophyll fluorescence, light-harvesting system and photoinhibition resistance of a zeaxanthin-accumulating mutant of *Arabidopsis thaliana*. *J Photochem Photobiol B, 34*(1), 87-94.

Terashima, I., Fujita, T., Inoue, T., Chow, W. S. and Oguchi, R. (2009). Green light drives leaf photosynthesis more efficiently than red light in strong white light: revisiting the enigmatic question of why leaves are green. *Plant Cell Physiol, 50*(4), 684-697. doi: 10.1093/pcp/pcp034

Terzer, M. and Stelling, J. (2008). Large-scale computation of elementary flux modes with bit pattern trees. *Bioinformatics, 24*(19), 2229-2235. doi: 10.1093/bioinformatics/btn401

Tohge, T., Ramos, M. S., Nunes-Nesi, A., Mutwil, M., Giavalisco, P., Steinhauser, D., Schellenberg, M., Willmitzer, L., Persson, S., Martinoia, E. and Fernie, A. R. (2011). Toward the storage metabolome: profiling the barley vacuole *Plant Physiol, 157*(3), 1469-1482. doi: 10.1104/pp.111.185710

Toriyama, K., Arimoto, Y., Uchimiya, H. and Hinata, K. (1988). Transgenic Rice Plants after Direct Gene-Transfer into Protoplasts. *Bio-Technology, 6*(9), 1072-1074. doi: Doi 10.1038/Nbt0988-1072

Trinh, C. T. (2012). Elucidating and reprogramming *Escherichia coli* metabolisms for obligate anaerobic n-butanol and isobutanol production. *Appl Microbiol Biotechnol, 95*(4), 1083-1094. doi: 10.1007/s00253-012-4197-7

Tschoep, H., Gibon, Y., Carillo, P., Armengaud, P., Szecowka, M., Nunes-Nesi, A., Fernie, A. R., Koehl, K. and Stitt, M. (2009). Adjustment of growth and central metabolism to a mild but sustained nitrogen-limitation in *Arabidopsis*. *Plant Cell Environ, 32*(3), 300-318. doi: 10.1111/j.1365-3040.2008.01921.x

Tzin, V. and Galili, G. (2010). The Biosynthetic Pathways for Shikimate and Aromatic Amino Acids in *Arabidopsis thaliana*. *Arabidopsis Book, 8*, e0132. doi: 10.1199/tab.0132

UNICEF. (2009). United Nations Children's Fund: Vitamin A deficiency: the challenge. Retrieved 03.04.2015, from http://www.childinfo.org/vitamina.htmL

Urbanczyk-Wochniak, E., Leisse, A., Roessner-Tunali, U., Lytovchenko, A., Reismeier, J., Willmitzer, L. and Fernie, A. R. (2003). Expression of a bacterial xylose isomerase in potato tubers results in an altered hexose composition and a consequent induction of metabolism. *Plant Cell Physiol, 44*(12), 1359-1367.

van Duuren, J. B., Puchalka, J., Mars, A. E., Bucker, R., Eggink, G., Wittmann, C. and Dos Santos, V. A. (2013). Reconciling *in vivo* and *in silico* key biological parameters of *Pseudomonas putida* KT2440 during growth on glucose under carbon-limited condition. *BMC Biotechnol, 13*, 93. doi: 10.1186/1472-6750-13-93

Van Norman, J. M. and Benfey, P. N. (2009). *Arabidopsis thaliana* as a model organism in systems biology. *Wiley Interdiscip Rev Syst Biol Med, 1*(3), 372-379. doi: 10.1002/wsbm.25

Vega, T., Breccia, G., Gil, M., Zorzoli, R., Picardi, L. and Nestares, G. (2012). Acetohydroxyacid synthase (AHAS) *in vivo* assay for screening imidazolinone-resistance in sunflower (*Helianthus annuus* L.). *Plant Physiol Biochem, 61*, 103-107. doi: 10.1016/j.plaphy.2012.09.013

Visser, W. F., van Roermund, C. W., Ijlst, L., Waterham, H. R. and Wanders, R. J. (2007). Metabolite transport across the peroxisomal membrane. *Biochem J, 401*(2), 365-375. doi: 10.1042/BJ20061352

von Caemmerer, S., Quick, W. P. and Furbank, R. T. (2012). The development of C₄ rice: current progress and future challenges. *Science, 336*(6089), 1671-1672. doi: 10.1126/science.1220177

von Kamp, A. and Schuster, S. (2006). Metatool 5.0: fast and flexible elementary modes analysis. *Bioinformatics, 22*(15), 1930-1931. doi: 10.1093/bioinformatics/btl267

Voss, I., Sunil, B., Scheibe, R. and Raghavendra, A. S. (2013). Emerging concept for the role of photorespiration as an important part of abiotic stress response. *Plant Biol (Stuttg), 15*(4), 713-722. doi: 10.1111/j.1438-8677.2012.00710.x

Wagner, C. (2004). Nullspace approach to determine the elementary modes of chemical reaction systems. *Journal of Physical Chemistry B, 108*(7), 2425-2431. doi: Doi 10.1021/Jp034523f

Wakasa, K., Hasegawa, H., Nemoto, H., Matsuda, F., Miyazawa, H., Tozawa, Y., Morino, K., Komatsu, A., Yamada, T., Terakawa, T. and Miyagawa, H. (2006). High-level tryptophan accumulation in seeds of transgenic rice and its limited effects on agronomic traits and seed metabolite profile. *J Exp Bot, 57*(12), 3069-3078. doi: 10.1093/jxb/erl068

Walk, T. (2007) Means and methods for analyzing a sample by means of chromotography-mass spectrometry. International Patent WO 2007/012643.

Wang, L., Birol, I. and Hatzimanikatis, V. (2004). Metabolic control analysis under uncertainty: framework development and case studies. *Biophys J, 87*(6), 3750-3763. doi: 10.1529/biophysj.104.048090

Weber, A. P. and Linka, N. (2011). Connecting the plastid: transporters of the plastid envelope and their role in linking plastidial with cytosolic metabolism. *Annu Rev Plant Biol, 62*, 53-77. doi: 10.1146/annurev-arplant-042110-103903

Weitzel, M., Nöh, K., Dalman, T., Niedenfuhr, S., Stute, B. and Wiechert, W. (2013). 13CFLUX2--high-performance software suite for ^{13}C-metabolic flux analysis. *Bioinformatics, 29*(1), 143-145. doi: 10.1093/bioinformatics/bts646

WHO. (2015). World Health Organization. Retrieved 03.04.2015, from http://www.who.int/nutrition/topics/vad/en/

Williams, T. C., Poolman, M. G., Howden, A. J., Schwarzlander, M., Fell, D. A., Ratcliffe, R. G. and Sweetlove, L. J. (2010). A genome-scale metabolic model accurately predicts fluxes in central carbon metabolism under stress conditions. *Plant Physiol, 154*(1), 311-323. doi: 10.1104/pp.110.158535

Wirth, J., Poletti, S., Aeschlimann, B., Yakandawala, N., Drosse, B., Osorio, S., Tohge, T., Fernie, A. R., Günther, D., Gruissem, W. and Sautter, C. (2009). Rice endosperm iron biofortification by targeted and synergistic action of nicotianamine synthase and ferritin. *Plant Biotechnol J,* *7*(7), 631-644. doi: 10.1111/j.1467-7652.2009.00430.x

Wirtz, M. and Hell, R. (2006). Functional analysis of the cysteine synthase protein complex from plants: structural, biochemical and regulatory properties. *J Plant Physiol, 163*(3), 273-286. doi: 10.1016/j.jplph.2005.11.013

Wittmann, C. (2002). Metabolic flux analysis using mass spectrometry. *Adv Biochem Eng Biotechnol,* *74*, 39-64.

Wittmann, C. (2007). Fluxome analysis using GC-MS. *Microb Cell Fact, 6*, 6. doi: 10.1186/1475-2859-6-6

Xie, T., Qiu, Q., Zhang, W., Ning, T., Yang, W., Zheng, C., Wang, C., Zhu, Y. and Yang, D. (2008). A biologically active rhIGF-1 fusion accumulated in transgenic rice seeds can reduce blood glucose in diabetic mice via oral delivery. *Peptides, 29*(11), 1862-1870. doi: 10.1016/j.peptides.2008.07.014

Xu, Y., McCouch, S. R. and Zhang, Q. (2005). How can we use genomics to improve cereals with rice as a reference genome? *Plant Mol Biol, 59*(1), 7-26. doi: 10.1007/s11103-004-4681-2

Yamamoto, T., Yonemaru, J. and Yano, M. (2009). Towards the understanding of complex traits in rice: substantially or superficially? *DNA Res, 16*(3), 141-154. doi: 10.1093/dnares/dsp006

Yan, H. and Jiang, J. (2007). Rice as a model for centromere and heterochromatin research. *Chromosome Res, 15*(1), 77-84. doi: 10.1007/s10577-006-1104-z

Yan, L. and Kerr, P. S. (2002). Genetically engineered crops: their potential use for improvement of human nutrition. *Nutr Rev, 60*(5 Pt 1), 135-141.

Yang, C., Li, D., Mao, D., Liu, X., Ji, C., Li, X., Zhao, X., Cheng, Z., Chen, C. and Zhu, L. (2013). Overexpression of microRNA319 impacts leaf morphogenesis and leads to enhanced cold tolerance in rice (*Oryza sativa* L.). *Plant Cell Environ, 36*(12), 2207-2218. doi: 10.1111/pce.12130

Yang, L., Hirose, S., Suzuki, K., Hiroi, T. and Takaiwa, F. (2012). Expression of hypoallergenic Der f 2 derivatives with altered intramolecular disulphide bonds induces the formation of novel ER-derived protein bodies in transgenic rice seeds. *J Exp Bot, 63*(8), 2947-2959. doi: 10.1093/jxb/ers006

Yin, X. and Struik, P. C. (2008). Applying modelling experiences from the past to shape crop systems biology: the need to converge crop physiology and functional genomics. *New Phytol, 179*(3), 629-642. doi: 10.1111/j.1469-8137.2008.02424.x

Yoo, H., Antoniewicz, M. R., Stephanopoulos, G. and Kelleher, J. K. (2008). Quantifying reductive carboxylation flux of glutamine to lipid in a brown adipocyte cell line. *J Biol Chem, 283*(30), 20621-20627. doi: 10.1074/jbc.M706494200

Young, J. D. (2014). INCA: a computational platform for isotopically non-stationary metabolic flux analysis. *Bioinformatics, 30*(9), 1333-1335. doi: 10.1093/bioinformatics/btu015

Young, J. D., Shastri, A. A., Stephanopoulos, G. and Morgan, J. A. (2011). Mapping photoautotrophic metabolism with isotopically nonstationary ^{13}C flux analysis. *Metab Eng, 13*(6), 656-665. doi: 10.1016/j.ymben.2011.08.002

Young, J. D., Walther, J. L., Antoniewicz, M. R., Yoo, H. and Stephanopoulos, G. (2008). An elementary metabolite unit (EMU) based method of isotopically nonstationary flux analysis. *Biotechnol Bioeng*, 99(3), 686-699. doi: 10.1002/bit.21632

Yu, W., Yau, Y. Y. and Birchler, J. A. (2015). Plant artificial chromosome technology and its potential application in genetic engineering. *Plant Biotechnol J*. doi: 10.1111/pbi.12466

Yu, X. Z. and Zhang, F. Z. (2013). Effects of exogenous thiocyanate on mineral nutrients, antioxidative responses and free amino acids in rice seedlings. *Ecotoxicology, 22*(4), 752-760. doi: 10.1007/s10646-013-1069-6

Yuan, J., Bennett, B. D. and Rabinowitz, J. D. (2008). Kinetic flux profiling for quantitation of cellular metabolic fluxes. *Nat Protoc, 3*(8), 1328-1340. doi: 10.1038/nprot.2008.131

Yuan, L. and Grötewold, E. (2015). Metabolic engineering to enhance the value of plants as green factories. *Metab Eng, 27*, 83-91. doi: 10.1016/j.ymben.2014.11.005

Zabaleta, E., Martin, M. V. and Braun, H. P. (2012). A basal carbon concentrating mechanism in plants? *Plant Sci, 187*, 97-104. doi: 10.1016/j.plantsci.2012.02.001

Zabalza, A., Orcaray, L., Gaston, S. and Royuela, M. (2004). Carbohydrate accumulation in leaves of plants treated with the herbicide chlorsulfuron or imazethapyr is due to a decrease in sink strength. *J Agric Food Chem, 52*(25), 7601-7606. doi: 10.1021/jf0486996

Zablackis, E., Huang, J., Muller, B., Darvill, A. G. and Albersheim, P. (1995). Characterization of the cell-wall polysaccharides of *Arabidopsis thaliana* leaves. *Plant Physiol, 107*(4), 1129-1138.

Zamboni, N., Fischer, E. and Sauer, U. (2005). FiatFlux--a software for metabolic flux analysis from ^{13}C-glucose experiments. *BMC Bioinformatics, 6*, 209. doi: 10.1186/1471-2105-6-209

Zanghellini, J., Ruckerbauer, D. E., Hanscho, M. and Jungreuthmayer, C. (2013). Elementary flux modes in a nutshell: properties, calculation and applications. *Biotechnol J, 8*(9), 1009-1016. doi: 10.1002/biot.201200269

Zhang, H. M., Yang, H., Rech, E. L., Golds, T. J., Davis, A. S., Mulligan, B. J., Cocking, E. C. and Davey, M. R. (1988). Transgenic rice plants produced by electroporation-mediated plasmid uptake into protoplasts. *Plant Cell Rep, 7*(6), 379-384. doi: 10.1007/BF00269517

Zhang, W. and Wu, R. (1988). Efficient regeneration of transgenic plants from rice protoplasts and correctly regulated expression of the foreign gene in the plants. *Theor Appl Genet, 76*(6), 835-840. doi: 10.1007/BF00273668

Zhang, X., Yuan, Z., Duan, Q., Zhu, H., Yu, H. and Wang, Q. (2009). Mucosal immunity in mice induced by orally administered transgenic rice. *Vaccine, 27*(10), 1596-1600. doi: 10.1016/j.vaccine.2008.12.042

Zhao, J. and Last, R. L. (1995). Immunological characterization and chloroplast localization of the tryptophan biosynthetic enzymes of the flowering plant *Arabidopsis thaliana*. *J Biol Chem, 270*(11), 6081-6087.

Zhu, X. G., Long, S. P. and Ort, D. R. (2010). Improving photosynthetic efficiency for greater yield. *Annu Rev Plant Biol, 61*, 235-261. doi: 10.1146/annurev-arplant-042809-112206

10. Appendix

10.1. Biomass composition of *A. thaliana* leaves and *O. sativa* seedling shoots

Biomass constitutes of many different building blocks, ranging from lipids, proteins and cell wall parts to steroids, nucleotides and organic acids. The content of the major biomass components of *A. thaliana* leaves and the overall composition used for elementary flux mode calculations, as well as their corresponding sources, are given in Table 10-1. For ^{13}C-INST-MFA calculations of *O. sativa* seedling shoots, several concentrations of biomass constituents were measured. These results, replenished with specific literature values, can be found in Table 10-2.

Table 10-1: Concentrations of all biomass components used in the elementary flux mode simulations of *A. thaliana* leaves accompanied by their respective source(s).

	Amount
	[mg (gDW)$^{-1}$]
Carbohydrates	
• monosaccharides	
○ glucose [A]	9.57
○ fructose [A]	1.60
○ fucose [B]	0.73
○ rhamnose [B]	0.84
○ arabinose [B]	0.53
○ mannose [B]	0.46
○ galactose [B]	6.99
○ xylose [B]	0.33
○ inositol [C]	3.20
• disaccharides	
○ sucrose [A]	16.02
• polysaccharides	
○ starch [A]	90.70

Cell wall
- cellulose [B]   46.74
- hemicellulose [D]
  - xyloglucan   66.77
  - glucuronoarabinoxylan   13.35
- pectin [D]   140.22
- lignin [E]
  - 4-coumaryl-alcohol   1.38
  - confideryl-alcohol   88.37
  - sinapyl-alcohol   48.33

- soluble polymers [B]
  - rhamnose   0.17
  - fucose   0.41
  - arabinose   2.22
  - xylose   0.49
  - mannose   0.60
  - galactose   4.69
  - glucose   0.15

Lipids
- glycerol [F]   0.40
- fatty acids [G,H]
  - C16:0   5.46
  - C16:1   1.41
  - C16:2   0.32
  - C16:3   4.48
  - C18:0   0.48
  - C18:1   1.24
  - C18:2   5.85
  - C18:3   19.22

Steroids [I]
- sitosterol   2.06
- stigmasterol   0.10

Proteins/Amino Acids [A]
- alanine   4.34
- arginine   5.79
- asparagine   12.28
- aspartate   36.25
- cysteine [J]   32.27
- glutamate   87.65
- glutamine   65.60
- glycine   9.01
- histidine   0.52
- isoleucine   0.46
- leucine   0.73
- lysine   0.41
- methionine   0.40
- phenylalanine   2.04
- proline   7.46
- serine   46.28

- threonine   8.77
- tryptophan   0.43
- tyrosine   0.29
- valine   1.54

Nucleotides
- DNA [K]
  - dATP   1.95
  - dTTP   1.95
  - dCTP   1.10
  - dGTP   1.10

- RNA [L]
  - ATP   0.22
  - UTP   0.22
  - CTP   0.13
  - GTP   0.13

Porphyrines
- chlorophyll [M]
  - chlorophyll a   3.81
  - chlorophyll b   7.78
- carotenoids [N]
  - β-carotene   0.49
  - zeaxanthin   0.06
  - lutein   1.07
  - antheraxanthin   0.02
  - violaxanthin   0.33
  - neoxanthin   0.26

Organic acids [A]
- malate   6.76
- fumarate   7.84

References: [A] Tschoep et al. (2009) [B] Reiter et al. (1997) [C] Saxena et al. (2013) [D] Zablackis et al. (1995) [E] Herrero et al. (2013) [F] Fan et al. (2013) [G] Stahl et al. (2004) [H] Shen et al. (2010) [I] Arnqvist et al. (2008) [J] de Oliveira Dal'Molin et al. (2010a) [K] Suzuki et al. (2004) [L] Murray and Thompson (1980) [M] Nowicka et al. (2009) [N] Tardy and Havaux (1996).

Table 10-2: Concentrations of all biomass components used in the flux calculations of rice seedling shoots, accompanied by their respective method or literature source(s).

	Amount [mg (gDW)$^{-1}$]		
	Wild type	**DMSO**	**Imazapyr**
amino acids [A,B]			
alanine	19.3	20.3	20.4
arginine	24.7	19.8	19.7
asparagine	30.7	33.2	32.9
aspartate	-	-	-
cysteine	2.1	1.9	4.6
glutamate	46.5	46.1	49.5
glutamine	-	-	-
glycine	17.6	18.1	17.4
histidine	7.5	7.5	7.4
isoleucine	13.4	13.8	13.5
leucine	26.9	27.6	26.9
lysine	18.0	19.2	18.4
methionine	5.0	5.0	4.9
phenylalanine	81.2	81.3	81.0
proline	14.8	15.2	14.6
serine	15.7	15.9	14.8
threonine	15.5	15.7	15.6
tryptophan [C]	0.03	0.03	0.03
tyrosine	52.4	52.4	49.5
valine	18.1	18.8	18.0
hemicellulose [D]			
xylose	53.6	53.6	53.6
galactose	5.6	5.6	5.6
arabinose	14.7	14.7	14.7
glucose	124.3	124.3	124.3
mannose	3.9	3.9	3.9
rhamnose	0.9	0.9	0.9
lignin [D]	42	42	42
cellulose [D]	52.5	52.5	52.5
starch [E]	6.1	4.1	7.2

soluble sugars [F]	66.7	61.5	70.1
sucrose	60.4	56.0	63.5
glucose	2.6	3.0	2.9
fructose	3.7	3.6	3.7
pigments [G]			
chlorophyll	21.9	21.9	21.9
carotenoids	0.7	0.7	0.7
nucleotides [H,I]			
RNA	13.3	13.3	13.3
DNA	8.2	8.2	8.2
lipids [J]	19.2	19.7	20.6
fatty acids (16.0)	2.5	2.7	2.8
fatty acids (18:0)	16.3	16.6	17.4
fatty acids (20:0)	0.3	0.4	0.4
organic acids	12.6	13.3	13.0
ascorbic acid [K]	6.0	6.0	6.0
citric acid [L]	1.1	1.5	1.4
succinic acid [L]	0.2	0.2	0.1
pyruvic acid [M]	0.6	0.5	0.5
2-oxoglutaric acid [M]	0.3	0.3	0.3
malic acid [L]	4.6	4.9	4.8

References: [A] C/N combustion [B] quantitative LC-MS/MS of amino acids [C] Ishihara et al. (2008) [D] Sumiyoshi et al. (2013) [E] Quantitative starch measurement [F] Quantitative free sugar measurement [G] Panda and Sarkar (2013) [H] Suzuki et al. (2001) [I] Murray and Thompson (1980) [J] Quantitative FAME measurement [K] Chao et al. (2010) [L] Quantitative GC-MS analysis [M] Quantitative LC-MS/MS analysis

10.2. Anabolic precursor demand for biomass synthesis of *A. thaliana* leaves and

O. sativa seedling shoots

Most pathways for synthesis of biomass building blocks are linear and can therefore be summarized into a single lumped biomass equation, depending only on a handful of precursors from central carbon metabolism. Information on the organism-specific pathway availability and stoichiometry can be found in the databases specified in the material and methods chapter. The anabolic precursor demand used for the modeling in this work are presented in Tables 10-3 and 10-4 for *A. thaliana* leaves and *O. sativa* seedlings, respectively.

Table 10-3 (next page): Precursor demand for synthesis of the different building blocks constituting *A. thaliana* leaf biomass.

	Protein	alanine	arginine	asparagine	aspartate	cysteine	glutamate	glutamine	glycine	histidine	isoleucine	leucine	lysine
SER[c]													
GLY[c]									1				
STA[p]													
MAL[c]													
FUM[m]													
G6P[c]													
F6P[c]													
ATP[m]			3				2	3					1
ATP[c]		2	4	4	2								
ATP[p]				1		4			2	6	4	2	3
GAP[p]													
NAD[c]													
NAD[p]						1				2	-1	1	
NADPH[m]			-1				1	1					
NADPH[c]			1	1	1	5				1			
NADPH[p]		1								1	3	2	4
Pi[m]			-2				-2	-2					-1
Pi[p]				-1		-5			-2	-8	-2	-2	-3
Pi[c]		-2	-4	-4	-2								
PYR[p]										1	2	1	
3PG[p]						1							
OAA[c]				1	1					1			1
E4P[p]													
DHAP[p]													
R5P[p]										1			
AKG[m]			1				1	1					
PEP[p]													
CO2[p]			1							1	-1	-2	-1
PYR[c]		1											
AcCoA[p]												1	

methionine		1						1	1	-1	-2	-6		5	4		1	2		7	2								
phenylalanine		-1	2				1					-6		2						3									
proline				1							-3				2	1					3	1							
serine												-2								2									1
threonine								1				-4		1	1		1			4									
tryptophan		-1	2		1		1		1	-1		-9		2			1		-1	6									
tyrosine		-1	2				1					-6		2			-1			3									
valine	2	-1										-2		2						2									
Carbohydrates																													
glucose											-2										1			1					
fructose																					1		1						
fucose											-2				1						1		1						
inositol											-1										1			1					
rhamnose											-2				1						1			1					
arabinose		-1									-2							2			1			1					

mannose		-2			1		1
galactose		-2			1		1
xylose	-1	-2		2	1		1
sucrose		-3			1	1	1
starch								1
Cell wall							
cellulose		-2			1		1
rhamnose		-2	1		1		1
fucose		-2	1		1	1	
arabinose	-1	-2		2	1		1
xylose	-1	-2		2	1	1	
mannose		-2			1		1
galactose		-2			1	1	1
glucose		-2			1		1
xyloglucan	-3	-14	1	6	7	1	6

glucuronoarabinoxylan	-2		4	-4			2	2
pectin	1			-2			1	1
Lignin								
4coumarylalcohol	-1	2	1	-8	5			
confiderylalcohol		2	1	-8	6			
sinapylalcohol	1	2	1	-8	7			
Porphyrines								
chlorophyll a	-9	8	4	-31	36		4	29
chlorophyll b	-9	8	4	-31	38		4	29
betacarotene	-8		8	-28	22		8	24
zeaxanthin	-8		8	-28	22	-2	8	24
lutein	-8		8	-28	24		8	24
antheraxanthin	-8		8	-28	23	-2	8	24
violaxanthin	-8		8	-28	24	-2	8	24
neoxanthin	-8		8	-28	24	-2	8	24

Nucleotides									
ATP		2	1		1		2	2	13
UTP			1	1			1		5
CTP			1	1			1		6
GTP		2	1		1		2	3	16
dATP		2	1		1		3	2	13
dTTP		1	1	1			3		8
dCTP			1	1			2		6
dGTP		2	1		1		3	3	16
Lipids									
C16:0	8					-8	7	-7	8
C16:1	8					-8	8	-7	8
C16:2	8					-8	8	-7	8
C16:3	8					-8	9	-7	8
C18:0	9					-9	8	-8	9

C18:1	9				-9	9	-8		9	
C18:2	9				-9	9	-8		9	
C18:3	9				-9	10	-8		9	
Glycerol			1			1				
Organic acids										
malate										1
fumarate										1
Stereoids										
sitosterol		-7		6	-18	25	-4	6	18	
stigmasterol		-7		6	-18	26	-4	6	18	

Table 10-4 (next page): Precursor demand for synthesis of the different building blocks constituting *O. sativa* seedling biomass, excluding the synthesis of amino acids and fatty acids, as these are accounted for by individual biosynthesis reactions in the metabolic network (Table 10-7).

| | Hemicellulose | | | | | | Lignin | Cellulose | hexose-phosphate | Soluble sugar |
	xylose	galactose	arabinose	glucose	mannose	rhamnose				sucrose
phenylalanine							1			
starch.p										
sucrose.cp										1
FRC.cp										
GLC.cp				1						
SUCC.m										
CIT.m										
MAL.m										
FUM.m										
G6P.cp		1	1			1			1	
F6P.cp					1					
GAP.cp										
3PG.cp										
OAA.c										
E4P.p										
P5P.p	1									
AKG.m										
PEP.cp										
CO2.p										
PYR.cp										
ACCOA.p										

glucose				1		
fructose					1	1
Starch						
starch						1
Porphyrines						
chlorophyll	4	-9	8	4		
carotinoide	8	-8		8		
Nucleotides						
RNA	2	2	3	1	-2	
DNA	2	2	3	1	-2	
Organic acids						
ascorbic acid				1		
citric acid					1	
isocitric acid					1	

succinic acid				1
pyruvic acid	1			
2oxoglutaric acid		1		
malic acid			1	

10.3. Stoichiometric reaction network of *A. thaliana* leaves

The reactions of the compartmentalized central carbon metabolism of *A. thaliana* leaves, used for elementary flux mode analysis, as well as their corresponding stoichiometry, are given in Table 10-5. A list of abbreviated metabolite names can be found in Appendix 10.5.

Table 10-5: Stoichiometric reaction network for *Arabidopsis thaliana* leaf metabolism. Stoichiometric factors of the biomass equation are expressed as mmol (gDW)$^{-1}$.

Reaction number	Reaction
	In silico transport reactions
1	'BM[c] -->'
2	'--> CO2EX[p]'
3	'--> STA[p]'
4	'CO2[cel]-->'
5	' <==> Pi[v]'
	Biomass Synthesis
6	'(1.238) ACCOA[p] + (0.075) PYR[c] + (0.360) CO2[cel] + (0.820) PEP[p] + (1.245) AKG[m] + (0.019) R5P[p] + (0.004) DHAP[p] + (0.410) E4P[p] + (0.455) OAA[c] + 0.278 3PG[p] + (0.126) PYR[p] + (2.206) NADPH[p] + (4.513) NADPH[c] + (1.143) NADPH[m] + (0.899) NADH[p] + (0.619) NAD[c] + (0.114) GAP[p] + (0.132) F6P[c] + (1.703) G6P[c] + (0.2652) STA[p] + (0.068) FUM[m] + (0.050) MAL[c] + 0.440 SER[c] + 0.120 GLY[c] + (4.652) ATP[p] + (6.013) ATP[c] + (2.705) ATP[m] --> BM[c] + (2.206) NADP[p] + (4.513) NADP[c] + (1.143) NADP[m] + (0.899) NAD[p] + (0.619) NADH[c] + (4.652) ADP[p] + (6.013) ADP[c] + (2.705) ADP[m] + (4.933) Pi[c] + (8.023) Pi[p] + (2.158) Pi[m]'
	Plastidic Metabolism
7	'G6P[p] <==> F6P[p]'
8	'F6P[p] + ATP[p] --> FBP[p] + ADP[p]'
9	'FBP[p] --> F6P[p] + Pi[p]'
10	'FBP[p] <==> DHAP[p] + GAP[p]'
11	'DHAP[p] <==> GAP[p]'
12	'GAP[p] + NADP[p] + Pi[p] <==> NADPH[p] + 13bPG[p]'
13	'13bPG[p] + ADP[p] <==> ATP[p] + 3PG[p]'
14	'3PG[p] <==> 2PG[p]'
15	'2PG[p] <==> PEP[p]'
16	'PEP[p] + ADP[p] --> PYR[p] + ATP[p]'
17	'PYR[p] + ATP[p] --> PEP[p] + AMP[p] + Pi[p]'
18	'AMP[p] + ATP[p] --> 2 ADP[p]'
19	'PYR[p] + E2Pr-lip[p] --> E2Pr-acet-lip[p] + CO2[cel]'
20	'E2Pr-acet-lip[p] --> ACCOA[p] + E2Pr-2hyd-lip[p]'
21	'E2Pr-2hyd-lip[p] + NAD[p] --> E2Pr-lip[p] + NADH[p]'
22	'MAL[p] + NADP[p] --> NADPH[p] + CO2[cel] + PYR[p]'

23	'G6P[p] + NADP[p] --> NADPH[p] + 6PGL[p]'
24	'6PGL[p] --> 6PG[p]'
25	'6PG[p] + NADP[p] --> NADPH[p] + RU5P[p] + CO2[cel]'
26	'R5P[p] <==> RU5P[p]'
27	'RU5P[p] <==> XU5P[p]'
28	'GAP[p] + S7P[p] <==> E4P[p] + F6P[p]'
29	'E4P[p] + XU5P[p] <==> F6P[p] + GAP[p]'
30	'S7P[p] + GAP[p] <==> R5P[p] + XU5P[p]'
31	'RU5P[p] + ATP[p] --> RBP[p] + ADP[p]'
32	'CO2EX[p] + RBP[p] --> (2) 3PG[p]'
33	'RBP[p] --> 2PGO[p] + 3PG[p]'
34	'2PGO[p] --> GLYCO[p] + Pi[p]'
35	'DHAP[p] + E4P[p] --> SBP[p]'
36	'SBP[p] --> S7P[p] + Pi[p]'
37	'GLYCER[p] + ATP[p] --> 3PG[p] + ADP[p]'
38	'MAL[p] + NAD[p] <==> OAA[p] + NADH[p]'
39	'MAL[p] + NADP[p] <==> OAA[p] + NADPH[p]'
40	'STA[p] --> MALT[p]'
41	'G6P[p] --> aG6P[p]'
42	'aG6P[p] --> G1P[p]'
43	'G1P[p] + ATP[p] --> ADP-GLC[p] + 2 Pi[p]'
44	'2 ADP-GLC[p] --> 2 ADP[p] + STA[p]'

Cytosolic Metabolism

45	'G6P[c] <==> F6P[c]'
46	'F6P[c] + ATP[c] <==> FBP[c] + ADP[c]'
47	'FBP[c] <==> DHAP[c] + GAP[c]'
48	'GAP[c] <==> DHAP[c]'
49	'GAP[c] + NADP[c] --> 3PG[c] + NADPH[c]'
50	'GAP[c] + NAD[c] + Pi[c] <==> 13dPG[c] + NADH[c]'
51	'13dPG[c] + ADP[c] <==> ATP[c] + 3PG[c]'
52	'3PG[c] <==> 2PG[c]'
53	'2PG[c] <==> PEP[c]'
54	'PYR[c] + ATP[c] --> PEP[c] + AMP[c] + Pi[c]'
55	'AMP[c] + ATP[c] --> 2 ADP[c]'
56	'PEP[c] + ADP[c] --> PYR[c] + ATP[c]'
57	'G6P[c] + NADP[c] --> NADPH[c] + 6PGL[c]'
58	'6PGL[c] --> 6PG[c]'
59	'6PG[c] + NADP[c] --> NADPH[c] + RU5P[c] + CO2[cel]'
60	'RU5P[c] <==> XU5P[c]'
61	'RU5P[c] <==> R5P[c]'
62	'OAA[c] + ATP[c] --> CO2[cel] + PEP[c] + ADP[c]'
63	'PEP[c] + CO2[cel] --> OAA[c] + Pi[c]'
64	'CIT[c] + ATP[c] --> ACCOA[c] + OAA[c] + ADP[c] + Pi[c]'
65	'MAL[c] + NAD[c] <==> OAA[c] + NADH[c]'
66	'MALT[c] --> 2 GLC[c]'
67	'GLC[c] + ATP[c] --> ADP[c] + G6P[c]'

	Perixosomal metabolism
68	'GLYCO[pe] --> GLYOX[pe]'
69	'GLYOX[pe] + NADH[pe] --> GLY[pe] + NAD[pe]'
70	'SER[pe] + GLYOX[pe] --> GLY[pe] + HPYR[pe]'
71	'HPYR[pe] + NADH[pe] --> NAD[pe] + GLYCER[pe]'
72	'MAL[pe] + NAD[pe] <==> OAA[pe] + NADH[pe]'
	Mitochondrial metabolism
73	'PYR[m] + E2Pr-lip[m] --> E2Pr-acet-lip[m] + CO2[cel]'
74	'E2Pr-acet-lip[m] --> ACCOA[m] + E2Pr-2hyd-lip[m]'
75	'E2Pr-2hyd-lip[m] + NAD[m] --> E2Pr-lip[m] + NADH[m]'
76	'ACCOA[m] + OAA[m] --> CIT[m]'
77	'CIT[m] --> ACO[m]'
78	'ACO[m] --> ICIT[m]'
79	'ICIT[m] + NAD[m] <==> AKG[m] + CO2[cel] + NADH[m]'
80	'ICIT[m] + NADP[m] <==> AKG[m] + CO2[cel] + NADPH[m]'
81	'AKG[m] + NAD[m] --> SUCCCOA[m] + CO2[cel] + NADH[m]'
82	'SUCCCOA[m] + ADP[m] + Pi[m] --> SUCC[m] + ATP[m]'
83	'SUCC[m] + UQN[m] --> FUM[m] + UQL[m]'
84	'UQL[m] + NAD[m] <==> UQN[m] + NADH[m]'
85	'FUM[m] --> MAL[m]'
86	'MAL[m] + NAD[m] --> OAA[m] + NADH[m]'
87	'MAL[m] + NAD[m] --> CO2[cel] + NADH[m] + PYR[m]'
88	'GLY[m] + MTHF[m] --> THF[m] + SER[m]'
89	'GLY[m] + LP[m] <==> CO2[cel] + SADHLP[m]'
90	'SADHLP[m] + THF[m] <==> MTHF[m] + DLP[m]'
91	'DLP[m] + NAD[m] <==> NADH[m] + LP[m]'
	Transporters
92	'GLYCO[p] <==> GLYCO[c]'
93	'GLYCO[c] <==> GLYCO[pe]'
94	'GLY[pe] <==> GLY[c]'
95	'GLY[c] <==> GLY[m]'
96	'SER[m] <==> SER[c]'
97	'SER[c] <==> SER[pe]'
98	'GLYCER[pe] <==> GLYCER[c]'
99	'GLYCER[c] <==> GLYCER[p]'
100	'MALT[p] --> MALT[c]'
101	'G6P[p] + Pi[c] <==> G6P[c] + Pi[p]'
102	'XU5P[p] + Pi[c] <==> XU5P[c] + Pi[p]'
103	'3PG[p] + Pi[c] <==> 3PG[c] + Pi[p]'
104	'DHAP[c] + Pi[p] <==> DHAP[p] + Pi[c] '
105	'PEP[c] + Pi[p] <==> PEP[p] + Pi[c]'
106	'PYR[p] <==> PYR[c]'
107	'PYR[c] <==> PYR[m]'
108	'MAL[c] --> MAL[pe]'
109	'OAA[pe] --> OAA[c]'
110	'MAL[c] + Pi[m] <==> MAL[m] + Pi[c]'

111	'OAA[c] + MAL[m] <==> MAL[c] + OAA[m]'
112	'OAA[c] + MAL[p] <==> MAL[c] + OAA[p]'
113	'OAA[c] + CIT[m] <==> CIT[c] + OAA[m]'
114	'Pi[c] --> Pi[m]'
115	'Pi[c] <==> Pi[p]'
116	'Pi[v] <==> Pi[c]'
117	'ATP[m] + ADP[c] --> ADP[m] + ATP[c]'
118	'ATP[p] + ADP[c] <==> ADP[p] + ATP[c]'

Energy metabolism

119	'2 Hv[p] + PQN[p] --> PQL[p] + 2 H_nc[p]'
120	'PQL[p] + 2 PC_o[p] <==> PQN[p] + 2 PC_r[p] + 2 H_nc[p]'
121	'Hv[p] + PC_r[p] + FE_o[p] --> PC_o[p] + FE_r[p]'
122	'2 FE_r[p] + NADP[p] --> 2 FE_o[p] + NADPH[p] + 2 H_nc[p]'
123	'Hv[p] --> 2 H_c[p]'
124	' --> Hv[p]'
125	'2 H_c[p] + 12 H_nc[p] + 3 ADP[p] --> 3 ATP[p]'
126	'ATP[p] --> ADP[p] + ATP_maint[cel]'
127	'(2.4) ADP[m] + NADH[m] + 2.4 Pi[m] --> NAD[m] + (2.4) ATP[m]'
128	' ATP_maint[cel] --> '
129	'NADP[c] + NADH[c] <==> NADPH[c] + NAD[c]'

Table 10-6: Slightly adjusted stoichiometric reaction network for *Arabidopsis thaliana* leaf metabolism for subsequent *in silico* target prediction. Stoichiometric factors of the biomass equation are expressed as mmol (gDW)$^{-1}$.

Reaction number	Reaction
	***In silico* transport reactions**
1	'BM[c] -->'
2	'--> CO2EX[p]'
3	'--> STA[p]'
4	'CO2[p]-->'
optional	'L-aspartate[c] -->'
optional	'Proline[m] -->'
optional	'STA[p] -->'
optional	'SUCR[c] -->'
optional	'LGNCEL[c] -->'
optional	'IPPP[p] -->'
optional	'Cysteine[p] -->'
optional	'Methionine[p] -->'
optional	'Threonine[p] -->'
optional	'Lysine[p] -->'
optional	'Tryptophan[p] -->'
	Biomass Synthesis
5	'(1.238) ACCOA[p] + (0.075) PYR[c] + (0.358) CO2[p] + (0.820) PEP[p] + (1.245) AKG[m] + 0.016 R5P[p] + 0.004 DHAP[p] + (0.410) E4P[p] + (0.455) OAA[c] + (0.827) 3PG[c] + 0.009 3PG[p] + (0.115) PYR[p] + (8.687) NADPH[p] + (0.384) NAD[p] + (14.341) ATP[p] + (0.112) GAP[p] + (0.131) F6P[c] + (0.530) G6P[p] + (2.097) G6P[c] + (0.068) FUM[m] + (0.050) MAL[c] --> BM[c] + (8.687) NADP[p] + (0.384) NADH[p] + (14.341) ADP[p]'
	Plastidic Metabolism
6	'G6P[p] <--> F6P[p]'
7	'F6P[p] + ATP[p] --> FBP[p] + ADP[p]'
8	'FBP[p] --> F6P[p]'
9	'FBP[p] <==> DHAP[p] + GAP[p]'
10	'DHAP[p] <==> GAP[p]'
11	'GAP[p] + NADP[p] + ADP[p] <==> NADPH[p] + ATP[p] + 3PG[p]'
12	'3PG[p] <==> PEP[p]'
13	'PEP[p] + ADP[p] --> PYR[p] + ATP[p]'
14	'PYR[p] + (2) ATP[p] --> PEP[p] + (2) ADP[p]'
15	'PYR[p] + NAD[p] --> ACCOA[p] + NADH[p] + CO2[p]'
16	'MAL[c] + NADP[p] --> NADPH[p] + CO2[p] + PYR[p]'
17	'G6P[p] + (2) NADP[p] --> (2) NADPH[p] + RU5P[p] + CO2[p]'
18	'R5P[p] <==> RU5P[p]'
19	'RU5P[p] <==> XU5P[p]'
20	'GAP[p] + S7P[p] <==> E4P[p] + F6P[p]'
21	'E4P[p] + XU5P[p] <==> F6P[p] + GAP[p]'

22	'S7P[p] + GAP[p] <==> R5P[p] + XU5P[p]'
23	'RU5P[p] + ATP[p] --> RBP[p] + ADP[p]'
24	'CO2EX[p] + RBP[p] --> (2) 3PG[p]'
25	'RBP[p] --> GLYOX[c] + 3PG[p]'
26	'DHAP[p] + E4P[p] --> S7P[p]'
27	'STA[p] + 4 ATP[p] --> 4 ADP[p] + 2 G6P[p]'

Cytosolic Metabolism

28	'G6P[c] <--> F6P[c]'
29	'F6P[c] + ATP[p] <==> DHAP[c] + GAP[c] + ADP[p]'
30	'GAP[c] <==> DHAP[c]'
31	'GAP[c] + NADP[p] --> 3PG[c] + NADPH[p]'
32	'GAP[c] + NAD[p] + ADP[p] <==> ATP[p] + 3PG[c] + NADH[p]'
33	'3PG[c] <==> PEP[c]'
34	'PYR[c] + (2) ATP[p] --> PEP[c] + (2) ADP[p]'
35	'PEP[c] + ADP[p] --> PYR[c] + ATP[p]'
36	'G6P[c] + (2) NADP[p] --> (2) NADPH[p] + RU5P[c] + CO2[p]'
37	'RU5P[c] <==> XU5P[c]'
38	'RU5P[c] <==> R5P[c]'
39	'(2) GLYOX[c] + (3) ATP[p] + (2) NADH[p] --> 3PG[p] + CO2[p] + (3) ADP[p] + (2) NAD[p]'
40	'OAA[c] + ATP[p] --> CO2[p] + PEP[c] + ADP[p]'
41	'PEP[c] + CO2[p] --> OAA[c] '

Perixosomal metabolism

42	'GLYOX[c] + ACCOA[c] --> MAL[c]'
43	'MAL[c] + NAD[p] <==> OAA[c] + NADH[p]'
44	'OAA[c] + ACCOA[c] --> CIT[c] '
45	'CIT[c] --> GLYOX[c] + SUCC[m] '
46	'CIT[c] + ATP[p] --> ACCOA[c] + OAA[c] + ADP[p]'

Mitochondrial metabolism

47	'PYR[m] + NAD[p] + OAA[m] --> NADH[p] + CO2[p] + CIT[m]'
48	'CIT[m] + NAD[p] <==> AKG[m] + CO2[p] + NADH[p]'
49	'CIT[m] + NADP[p] <==> AKG[m] + CO2[p] + NADPH[p]'
50	'AKG[m] + NAD[p] + ADP[p] --> SUCC[m] + CO2[p] + ATP[p] + NADH[p]'
51	'SUCC[m] + NAD[p] --> FUM[m] + NADH[p]'
52	'FUM[m] --> MAL[m]'
53	'MAL[m] + NAD[p] --> OAA[m] + NADH[p]'
54	'MAL[m] + NAD[p] --> CO2[p] + NADH[p] + PYR[m]'

Transporters

55	'DHAP[c] <==> DHAP[p]'
56	'MAL[c] <==> MAL[m]'
57	'G6P[p] <==> G6P[c]'

58	'3PG[p] <==> 3PG[c]'
59	'PYR[p] <==> PYR[c]'
60	'PYR[c] <==> PYR[m]'
61	'PEP[c] <==> PEP[p]'
62	'XU5P[p] <==> XU5P[c]'
63	'OAA[c] + MAL[m] <==> MAL[c] + OAA[m]'
64	'OAA[c] + CIT[m] <==> CIT[c] + OAA[m]'

Energy metabolism

65	'(9) Hv[p] + (2) NADP[p] + (3) ADP[p] --> (2) NADPH[p] + (3) ATP[p]'
66	'ATP[p] --> ADP[p] + ATP_maint[p]'
67	'(2.4) ADP[p] + NADH[p] --> NAD[p] + (2.4) ATP[p]'
68	' --> Hv[p]'
69	' ATP_maint[p] --> '
70	'NADP[p] + NADH[p] <==> NADPH[p] + NAD[p]'

Product Synthesis

optional	'L-aspartate[c] + ATP[p] + (2) NADPH[p] --> Homoserine[p] + ADP[p] + (2) NADP[p]'
optional	'Homoserine[p] + ATP[p] --> ADP[p] + Threonine[p]'
optional	'Homoserine[p] + Cysteine[p] + (2) ATP[p] --> (2) ADP[p] + PYR[p] + Methionine[p]'
optional	'OAA[c] + NADPH[p] --> NADP[p] + L-aspartate[c]'
optional	'3PG[p] + NAD[p] + 5 NADPH[p] + 4 ATP[p] --> NADH[p] + 5 NADP[p] + 4 ADP[p] + Cysteine[p]'
optional	'AKG[m] + (3) NADPH[p] + ATP[p] --> Proline[m] + (3) NADP[p] + ADP[p]'
optional	'PYR[p] + GAP[p] + (2) NADPH[p] + (2) ATP[p] + (2) Ferox[p] --> IPPP[p] + (2) ADP[p] + (2) NADP[p] + CO2[p] + (2) Ferred[p]'
optional	'PYR[p] + GAP[p] + NADPH[p] + NADH[p] + (2) ATP[p] + (2) Ferox[p] --> IPPP[p] + (2) ADP[p] + NAD[p] + NADP[p] + CO2[p] + (2) Ferred[p]'
optional	'(3) ACCOA[p] + (2) NADPH[p] + (3) ATP[p] --> IPPP[p] + (3) ADP[p] + (2) NADP[p] + CO2[p]'
optional	'(2) Ferox[p] + NADPH[p] --> (2) Ferred[p] + NADP[p]'
optional	'ATP[p] + F6P[c] + G6P[c] --> SUCR[c] + ADP[p]'
optional	'(2) ATP[p] + (2) G6P[p] --> STA[p] + (2) ADP[p]'
optional	'(10.950) ATP[p] + (2.363) G6P[c] + (9.064) NADPH[p] + (1.363) E4P[p] + (2.726) PEP[p] --> LGNCEL[c] + (0.450) CO2[p] + (9.064) NADP[p] + (10.950) ADP[p]'
optional	'L-aspartate[c] + ATP[p] + (3) NADPH[p] + PYR[p] --> Lysine[p] + ADP[p] + (3) NADP[p] + CO2[p]'
optional	'2 PEP[p] + R5P[p] + E4P[p] + 3PG[p] + 2 NADPH[p] + NAD[p] + 2 ATP[p] -> Tryptophan[p] + CO2[p] + PYR[p] + 2 NADP[p] + NADH[p] + GAP[p] + 2 ADP[p]'

10.4. Isotopomer reaction network of *O. sativa* seedling shoots

The reactions of the large-scale isotopomer network of *O. sativa* seedling shoots, used for ^{13}C-INST-MFA, as well as their corresponding stoichiometry and atom transitions, are given in Tables 10-7 and 10-8. The former describes the finalized condensed network for the wild type rice seedling, whereas the latter details a fully compartmentalized model, used as starting point in the iterative development of the finalized network topology. A list of abbreviated metabolite names can be found in Appendix 10.5.

Table 10-7: Finalized isotopomer reaction network for *Oryza sativa* seedling metabolism. Stoichiometric factors of the biomass equation are expressed as mmol (gDW)$^{-1}$ and atom transitions are represented between brackets.

Reaction number	Reaction
	CO$_2$ metabolism
1	CO2in.p (a) -> CO2EX.p (a)
2	12CO2in.p (a) -> CO2in.p (a)
3	13CO2in.p (a) -> CO2in.p (a)
4	CO2.p (a) -> CO2EX.p (a)
5	CO2.p (a) -> CO2sink.s (a)
	Biomass Synthesis (untreated)
6	0.1181*PYR.cp + 0.1979*AKG.m + 0.4437*P5P.p + 0.1299*OAA.c + 0.0433*3PG.cp + 0.0218*F6P.cp + 0.3704*G6P.cp + 0.1083*GAP.cp + 0.0055*CIT.m + 0.0012*SUCC.m + 0.0342*MAL.m + 0.0206*FRC.cp + 0.7041*GLC.cp + 0.1765*sucrose.cp + 0.0377*starch.p + 0.2171*alanine.c + 0.1417*arginine.c + 0.2327*aspar.c + 0.0173*cysteine.p + 0.3163*glutamate.m + 0.2341*glycine.p + 0.0483*histidine.p + 0.1022*isoleucine.p + 0.2047*leucine.p + 0.1229*lysine.p + 0.0337*methionine.p + 0.7690*phenylalanine.p + 0.1282*proline.c + 0.1422*serine.p + 0.1297*threonine.p + 0.2891*tyrosine.p + 0.1548*valine.p + 0.0001*tryptophane.p + 0.0098*C16.p + 0.0573*C18.p + 0.001075*C20.p -> biomass
	Biomass Synthesis (DMSO-treated)
6	0.1170*PYR.cp + 0.1982*AKG.m + 0.4437*P5P.p + 0.1299*OAA.c + 0.0433*3PG.cp + 0.0218*F6P.cp + 0.3704*G6P.cp + 0.1083*GAP.cp + 0.0076*CIT.m + 0.0015*SUCC.m + 0.0362*MAL.m + 0.0202*FRC.cp + 0.7062*GLC.cp + 0.1603 sucrose.cp + 0.0253 starch.p + 0.2275*alanine.c + 0.1137*arginine.c + 0.2511*aspar.c + 0.0157*cysteine.p + 0.3135*glutamate.m + 0.2406*glycine.p + 0.0484*histidine.p + 0.1054*isoleucine.p + 0.2108*leucine.p + 0.1315*lysine.p + 0.0334*methionine.p + 0.7599*phenylalanine.p + 0.1322*proline.c + 0.1517*serine.p + 0.1320*threonine.p + 0.2894*tyrosine.p + 0.1603*valine.p + 0.0001*tryptophane.p + 0.0105 C16.p + 0.0584 C18.p + 0.0012 C20.p -> biomass

6	**Biomass Synthesis (imazapyr-treated)** 0.1174*PYR.cp + 0.1978*AKG.m + 0.4437*P5P.p + 0.1299*OAA.c + 0.0433*3PG.cp + 0.0218*F6P.cp + 0.3704*G6P.cp + 0.1083*GAP.cp + 0.0073*CIT.m + 0.0008*SUCC.m + 0.0354*MAL.m + 0.0204*FRC.cp + 0.7060*GLC.cp + 0.1856 sucrose.cp + 0.0443 starch.p + 0.2291*alanine.c + 0.1133*arginine.c + 0.2493*aspar.c + 0.038*cysteine.p + 0.3367*glutamate.m + 0.2319*glycine.p + 0.0475*histidine.p + 0.1028*isoleucine.p + 0.205*leucine.p + 0.1255*lysine.p + 0.0329*methionine.p + 0.7580*phenylalanine.p + 0.1270*proline.c + 0.1500*serine.p + 0.1314*threonine.p + 0.2734*tyrosine.p + 0.1535*valine.p + 0.0001*tryptophane.p + 0.011 C16.p + 0.061 C18.p + 0.0011 C20.p -> biomass

	Embden-Meyerhof-Parnas pathway
7	G6P.cp (abcdef) <-> F6P.cp (abcdef)
8	FBP.cp (abcdef) <-> F6P.cp (abcdef)
9	FBP.cp (abcdef) <-> DHAP.cp (cba) + GAP.cp (def)
10	GAP.cp (abc) <-> DHAP.cp (abc)
11	GAP.cp (abc) <-> 3PG.cp (abc)
12	3PG.cp (abc) <-> 2PG.cp (abc)
13	2PG.cp (abc) <-> PEP.cp (abc)
14	PYR.cp (abc) <-> PEP.cp (abc)
15	PYR.cp (abc) -> ACCOA.p (bc) + CO2.p (a)

	Oxidative pentose phosphate pathway
16	G6P.cp (abcdef) -> 6PG.cp (abcdef)
17	6PG.cp (abcdef) -> P5P.p (bcdef) + CO2.p (a)

	Non-oxidative pentose phosphate pathway
18	GAP.cp (hij) + S7P.p (abcdefg) <-> E4P.p (defg) + F6P.p (abchij)
19	E4P.p (cdef) + P5P.p (abghi) <-> F6P.p (abcdef) + GAP.cp (ghi)
20	GAP.cp (hij) + S7P.p (abcdefg) <-> P5P.p (cdefg) + P5P.p (abhij)

	Calvin-Benson-Bassham cycle
21	P5P.p (abcde) -> RBP.p (abcde)
22	CO2EX.p (f) + RBP.p (abcde) -> 3PG.cp (cba) + 3PG.cp (fde)
23	DHAP.cp (cba) + E4P.p (defg) -> S7P.p (abcdefg)

	Photorespiration
24	RBP.p (abcde) -> GLYCO.pg (ba) + 3PG.cp (cde)
25	GLYCO.pg (ab) -> GLYOX.pg (ab)
26	GLYOX.pg (ab) + serine.p (cde) -> glycine.p (ab) + GLYCER.pg (cde)
27	GLYCER.pg (abc) -> 3PG.cp (abc)

	Amphibolic reactions
28	CO2.p (d) + PEP.cp (abc) -> OAA.c (abcd)

29	MAL.p (abcd) <-> PYR.cp (abc) + CO2.p (d)

Transporters

30	MAL.m (abcd) <-> MAL.c (abcd)
31	PYR.m (abc) <-> PYR.cp (abc)
32	OAA.m (abcd) + CIT.c (efghij) <-> OAA.c (abcd) + CIT.m (efghij)
33	MAL.m (abcd) + OAA.c (efgh) <-> MAL.c (abcd) + OAA.m (efgh)
34	MAL.c (abcd) <-> MAL.p (abcd)

Mitochondrial metabolism

35	PYR.m (abc) -> ACCOA.m (bc) + CO2.p (a)
36	OAA.m (abcd) + ACCOA.m (ef) -> CIT.m (dcbefa)
37	CIT.m (abcdef) <-> AKG.m (abcde) + CO2.p (f)
38	AKG.m (abcde) -> SUCC.m (bcde) + CO2.p (a)
39	SUCC.m (abcd) -> FUM.m (abcd)
40	FUM.m (abcd) -> MAL.m (abcd)
41	MAL.m (abcd) -> OAA.m (abcd)
42	MAL.m (abcd) -> CO2.p (d) + PYR.m (abc)

Starch synthesis

43	G6P.cp -> starch.p

Sugar metabolism

44	G6P.cp (abcdef) -> INO.cp (abcdef)
45	G6P.cp (abcdef) -> GLC.cp (abcdef)
46	GLC.cp (abcdef) -> G6P.cp (abcdef)
47	F6P.cp (abcdef) -> FRC.cp (abcdef)
48	G6P.cp (abcdef) + F6P.cp (ghijkl) -> sucrose.cp (fedcbahijklg)
49	sucrose.cp (abcdefghijkl) -> sucrose.r (abcdefghijkl)

Fatty acid metabolism

50	8*ACCOA.p -> C16.p
51	9*ACCOA.p -> C18.p
52	10*ACCOA.p -> C20.p

Amino acid metabolism

53	3PG.cp (abc) -> serine.p (abc)
54	3PG.cp (abc) -> cysteine.p (abc)
55	3PG.cp (abc) -> glycine.p (ab) + CO2.p (c)
56	GLYOX.c (ab) -> glycine.p (ab)
57	glycine.p (ab) -> MTHF.p (b) + CO2.p (a)
58	glycine.p (ab) + MTHF.p (c) -> serine.p (abc)
59	PYR.cp (abc) + glutamate.m (defgh) -> alanine.c (abc) + AKG.m (defgh)
60	PYR.cp (abc) + PYR.cp (def) + glutamate.m (ghijk) -> valine.p (abefc) + AKG.m (ghijk) + CO2.p (d)
61	PYR.cp (abc) + PYR.cp (def) + ACCOA.p (gh) + glutamate.m (ijklm) -> leucine.p (ghbefc) + CO2.p (a) + CO2.p (d) + AKG.m (ijklm)

62	OAA.c (abcd) + glutamate.m (efghi) -> aspar.c (abcd) + AKG.m (efghi)
63	OAA.c (abcd) -> aspar.c (abcd)
64	P5P.p (abcde) + CO2.p (f) + glutamate.m (ghijk) -> histidine.p (fabcde) + AKG.m (ghijk)
65	AKG.m (abcde) -> proline.c (abcde)
66	AKG.m (abcde) + glutamate.m (klmno) + OAA.c (fghi) + CO2.p (j) -> arginine.c (abcdej) + AKG.m
67	(klmno) + FUM.m (fghi)
68	OAA.c (abcd) + PYR.cp (efg) + glutamate.m (hijkl) -> lysine.p (abcdgf) + CO2.p (e) + AKG.m (hijkl)
69	OAA.c (abcd) -> threonine.p (abcd)
70	OAA.c (defg) + PYR.cp (abc) + glutamate.m (hijkl) -> isoleucine.p (debfgc) + CO2.p (a) + AKG.m
	(hijkl)
71	OAA.c (abcd) + 3PG.cp (efg) + CO2.p (h) -> methionine.p (abcdh) + PYR.cp (efg)
72	E4P.p (defg) + PEP.cp (abc) + PEP.cp (hij) + glutamate.m (klmno) -> tyrosine.p (hijbcdefg) + CO2.p
	(a) + AKG.m (klmno)
73	E4P.p (defg) + PEP.cp (abc) + PEP.cp (hij) + glutamate.m (klmno) -> phenylalanine.p (hijbcdefg) +
	CO2.p (a) + AKG.m (klmno)
74	AKG.m (abcde) -> glutamate.m (abcde)
75	E4P.p (defg) + PEP.cp (abc) + PEP.cp (hij) + P5P.p (klmno) + 3PG.cp (qrs) + glutamate.m (tuvwx) -
	> CO2.p (a) + tryptophane.p (bcdefgklsrq) + PYR.cp (hij) + GAP.cp (mno) + AKG.m (tuvwx)

Table 10-8: Initial fully compartmentalized isotopomer reaction network for *Oryza sativa* seedling metabolism. Stoichiometric factors of the biomass equation are expressed as mmol $(gDW)^{-1}$ and atom transitions are represented between brackets.

Reaction number	Reaction
	CO₂ metabolism
1	CO2in.p (a) -> CO2EX.p (a)
2	12CO2in.p (a) -> CO2in.p (a)
3	13CO2in.p (a) -> CO2in.p (a)
4	CO2.p (a) -> CO2EX.p (a)
5	CO2.p (a) -> CO2sink.s (a)
	Biomass Synthesis (untreated)
6	0.1181*PYR.p + 0.1979*AKG.m + 0.4437*P5P.p + 0.1299*OAA.c + 0.0433*3PG.c + 0.0218*F6P.c + 0.3704*G6P.c + 0.1083*GAP.p + 0.0055*CIT.m + 0.0012*SUCC.m + 0.0342*MAL.m + 0.0206*FRC.cp + 0.7041*GLC.cp + 0.1765*sucrose.cp + 0.0377*starch.p + 0.2171*alanine.c + 0.1417*arginine.c + 0.2327*aspar.c + 0.0173*cysteine.p + 0.3163*glutamate.m + 0.2341*glycine.p + 0.0483*histidine.p + 0.1022*isoleucine.p + 0.2047*leucine.p + 0.1229*lysine.p + 0.0337*methionine.p + 0.769*phenylalanine.p + 0.1282*proline.c + 0.0377*serine.p + 0.1297*threonine.p + 0.2891*tyrosine.p + 0.1548*valine.p + 0.0001*tryptophane.p + 0.0098*C16.p + 0.0573*C18.p + 0.001075*C20.p -> biomass
	Cytosolic Embden-Meyerhof-Parnas pathway
7	G6P.c (abcdef) <-> F6P.c (abcdef)
8	FBP.c (abcdef) <-> F6P.c (abcdef)
9	FBP.c (abcdef) <-> DHAP.c (cba) + GAP.c (def)
10	GAP.c (abc) <-> DHAP.c (abc)
11	GAP.c (abc) <-> 3PG.c (abc)
12	3PG.c (abc) <-> 2PG.c (abc)
13	2PG.c (abc) <-> PEP.c (abc)
14	PYR.c (abc) <-> PEP.c (abc)
	plastidic Embden-Meyerhof-Parnas pathway
15	G6P.p (abcdef) <-> F6P.p (abcdef)
16	FBP.p (abcdef) <-> F6P.p (abcdef)
17	FBP.p (abcdef) <-> DHAP.p (cba) + GAP.p (def)
18	GAP.p (abc) <-> DHAP.p (abc)
19	GAP.p (abc) <-> 3PG.p (abc)
20	3PG.p (abc) <-> 2PG.p (abc)
21	2PG.p (abc) <-> PEP.p (abc)
22	PYR.p (abc) <-> PEP.p (abc)
23	PYR.p (abc) -> ACCOA.p (bc) + CO2.p (a)

Oxidative pentose phosphate pathway

24	G6P.c (abcdef) -> 6PG.c (abcdef)
25	6PG.c (abcdef) -> P5P.c (bcdef) + CO2.p (a)
26	G6P.p (abcdef) -> 6PG.p (abcdef)
27	6PG.p (abcdef) -> P5P.p (bcdef) + CO2.p (a)

Non-oxidative pentose phosphate pathway

28	GAP.p (hij) + S7P.p (abcdefg) <-> E4P.p (defg) + F6P.p (abchij)
29	E4P.p (cdef) + P5P.p (abghi) <-> F6P.p (abcdef) + GAP.p (ghi)
30	GAP.p (hij) + S7P.p (abcdefg) <-> P5P.p (cdefg) + P5P.p (abhij)

Calvin-Benson-Bassham cycle

31	P5P.p (abcde) -> RBP.p (abcde)
32	CO2EX.p (f) + RBP.p (abcde) -> 3PG.p (cba) + 3PG.p (fde)
33	DHAP.p (cba) + E4P.p (defg) -> S7P.p (abcdefg)

Photorespiration

34	RBP.p (abcde) -> GLYCO.pg (ba) + 3PG.p (cde)
35	GLYCO.pg (ab) -> GLYOX.pg (ab)
36	GLYOX.pg (ab) + serine.p (cde) -> glycine.p (ab) + GLYCER.pg (cde)
37	GLYCER.pg (abc) -> 3PG.p (abc)

Amphibolic reactions

38	CO2.p (d) + PEP.c (abc) -> OAA.c (abcd)
39	MAL.p (abcd) <-> PYR.p (abc) + CO2.p (d)

Transporters

40	MAL.m (abcd) <-> MAL.c (abcd)
41	PYR.m (abc) <-> PYR.cp (abc)
42	OAA.m (abcd) + CIT.c (efghij) <-> OAA.c (abcd) + CIT.m (efghij)
43	MAL.m (abcd) + OAA.c (efgh) <-> MAL.c (abcd) + OAA.m (efgh)
44	MAL.c (abcd) <-> MAL.p (abcd)
45	'3PG.c (abc) <-> 3PG.p (abc)
46	DHAP.c (abc) <-> DHAP.p (abc)
47	PEP.c (abc) <-> PEP.p (abc)
48	PYR.c (abc) <-> PYR.p (abc)
49	G6P.c (abcdef) <-> G6P.p (abcdef)

Mitochondrial metabolism

50	PYR.m (abc) -> ACCOA.m (bc) + CO2.p (a)
51	OAA.m (abcd) + ACCOA.m (ef) -> CIT.m (dcbefa)
52	CIT.m (abcdef) <-> AKG.m (abcde) + CO2.p (f)

53	AKG.m (abcde) -> SUCC.m (bcde) + CO2.p (a)
54	SUCC.m (abcd) -> FUM.m (abcd)
55	FUM.m (abcd) -> MAL.m (abcd)
56	MAL.m (abcd) -> OAA.m (abcd)
57	MAL.m (abcd) -> CO2.p (d) + PYR.m (abc)

Starch synthesis

| 58 | G6P.p -> starch.p |

Sugar metabolism

59	G6P.c (abcdef) -> INO.c (abcdef)
60	G6P.c (abcdef) -> GLC.c (abcdef)
61	GLC.c (abcdef) -> G6P.c (abcdef)
62	F6P.c (abcdef) -> FRC.c (abcdef)
63	G6P.c (abcdef) + F6P.c (ghijkl) -> sucrose.c (fedcbahijklg)
64	sucrose.c (abcdefghijkl) -> sucrose.r (abcdefghijkl)

Fatty acid metabolism

65	8*ACCOA.p -> C16.p
66	9*ACCOA.p -> C18.p
67	10*ACCOA.p -> C20.p

Amino acid metabolism

68	3PG.p (abc) -> serine.p (abc)
69	3PG.p (abc) -> cysteine.p (abc)
70	3PG.p (abc) -> glycine.p (ab) + CO2.p (c)
71	GLYOX.g (ab) -> glycine.p (ab)
72	glycine.p (ab) -> zzMTHF.p (b) + CO2.p (a)
73	glycine.p (ab) + zzMTHF.p (c) -> serine.p (abc)
74	PYR.c (abc) + glutamate.m (defgh) -> alanine.c (abc) + AKG.m (defgh)
75	PYR.p (abc) + PYR.p (def) + glutamate.m (ghijk) -> valine.p (abefc) + AKG.m (ghijk) + CO2.p (d)
76	PYR.p (abc) + PYR.p (def) + ACCOA.p (gh) + glutamate.m (ijklm) -> leucine.p (ghbefc) + CO2.p (a) + CO2.p (d) + AKG.m (ijklm)
77	OAA.c (abcd) + glutamate.m (efghi) -> aspar.c (abcd) + AKG.m (efghi)
78	OAA.c (abcd) -> aspar.c (abcd)
79	P5P.p (abcde) + CO2.p (f) + glutamate.m (ghijk) -> histidine.p (fabcde) + AKG.m (ghijk)
80	AKG.m (abcde) -> proline.c (abcde)
81	AKG.m (abcde) + glutamate.m (klmno) + OAA.c (fghi) + CO2.p (j) -> arginine.c (abcdej) + AKG.m
82	(klmno) + FUM.m (fghi)
83	OAA.c (abcd) + PYR.p (efg) + glutamate.m (hijkl) -> lysine.p (abcdgf) + CO2.p (e) + AKG.m (hijkl)
84	OAA.c (abcd) -> threonine.p (abcd)
85	OAA.c (defg) + PYR.p (abc) + glutamate.m (hijkl) -> isoleucine.p (debfgc) + CO2.p (a) + AKG.m (hijkl)
86	OAA.c (abcd) + 3PG.p (efg) + CO2.p (h) -> methionine.p (abcdh) + PYR.p (efg)
87	E4P.p (defg) + PEP.p (abc) + PEP.p (hij) + glutamate.m (klmno) -> tyrosine.p (hijbcdefg) + CO2.p (a) + AKG.m (klmno)
88	E4P.p (defg) + PEP.p (abc) + PEP.p (hij) + glutamate.m (klmno) -> phenylalanine.p (hijbcdefg) + CO2.p (a) + AKG.m (klmno)
89	AKG.m (abcde) -> glutamate.m (abcde)

| 90 | E4P.p (defg) + PEP.p (abc) + PEP.p (hij) + P5P.p (klmno) + 3PG.p (qrs) + glutamate.m (tuvwx) -> CO2.p (a) + tryptophane.p (bcdefgklsrq) + PYR.p (hij) + GAP.p (mno) + AKG.m (tuvwx) |

10.5.   Abbreviations of metabolite names used in the network reconstructions

10.5.1. Abbreviations used in the *A. thaliana* stoichiometric network model

Table 10-9: Abbreviations of metabolite names used in the stoichiometric network model of *A. thaliana* from Table 10-5.

Abbreviation	metabolite name	Abbreviation	metabolite name
13bPG	1,3-bisphosphoglycerate	GLYOX	glyoxylate
2PG	2-phosphoglycerate	H_c	H⁺ produced through cyclic photophosphorylation
2PGO	2-phosphoglycolate	H_nc	H⁺ produced through non-cyclic photophosphorylation
3PG	3-phosphoglycerate	HPYR	hydroxypyruvate
6PG	gluconate 6-phosphate	Hv	photon
6PGL	6-phospho glucono-1,5-lactone	ICIT	isocitrate
ACCOA	acetyl-CoA	LP	lipoylprotein
ACO	cis-aconitate	MAL	malate
ADP	adenosine diphosphate	MALT	maltose
ADP-GLC	ADP-glucose	MTHF	5-methyltetrahydrofolate
aG6P	α-glucose 6-phosphate	NAD	nicotinamide adenine dinucleotide (oxidized)
AKG	α-ketoglutarate	NADH	nicotinamide adenine dinucleotide (reduced)

AMP	adenosine monophosphate	NADP	nicotinamide adenine dinucleotide phosphate (oxidized)
ATP	adenosine triphosphate	NADPH	nicotinamide adenine dinucleotide phosphate (reduced)
ATP_maint	maintenance ATP	OAA	oxaloacetate
BM	biomass	PC_o	plastocyanin (oxidized)
CIT	citrate	PC_r	plastocyanin (reduced)
CO2	internal CO_2	PEP	phosphoenolpyruvate
CO2EX	assimilated CO_2	Pi	phosphate
DHAP	dihydroxyacetone phosphate	PQL	plastoquinol
DLP	dihydrolipoylprotein	PQN	plastoquinone
E2Pr-2hyd-lip	[pyruvate dehydrogenase E2 protein]- dihydrolipoyl-L-lysine	PYR	pyruvate
E2Pr-acet-lip	[pyruvate dehydrogenase E2 protein]-S-acetyldihydrolipoyl-L-lysine	R5P	ribose 5-phosphate
E2Pr-lip	[pyruvate dehydrogenase E2 protein]-lipoyl-L-lysine	RBP	ribulose 1.5-bisphosphate
E4P	erythrose 4-phosphate	RU5P	ribulose 5-phosphate
F6P	fructose 6-phosphate	S7P	sedoheptulose 7-phosphate
FBP	fructose 1.6-bisphosphate	SADHLP	S-aminomethyl-dihydrolipoylprotein

FE_o	ferredoxin (oxidized)	SBP	sedoheptulose 1,7-bisphosphate
FE_r	ferredoxin (reduced)	SER	serine
FUM	fumarate	STA	starch
G1P	glucose 1-phosphate	SUCC	succinate
G6P	glucose 6-phosphate	SUCCCOA	succinyl-CoA
GAP	glyceraldehyde 3-phosphate	THF	tetrahydrofolate
GLC	glucose	UQL	ubiquinol
GLY	glycine	UQN	ubiquinone
GLYCER	glycerate	XU5P	xylulose 5-phosphate
GLYCO	glycolate		

10.5.2. Abbreviations used in the *O. sativa* isotopomer network model

Table 10-10: Abbreviations of metabolite names used in the isotopomeric network model of *O. sativa* from Table 10-7 and 10-8.

Abbreviation	metabolite name	Abbreviation	metabolite name
12CO2in	artificial $^{12}CO_2$ input	G6P	glucose 6-phosphate
13CO2in	artificial $^{13}CO_2$ input	GAP	glyceraldehyde 3-phosphate
2PG	2-phosphoglycerate	aspar	aspartate
3PG	3-phosphoglycerate	GLC	glucose
6PG	6-phosphoglyconate	GLL6P	6-phosphogluconolactone
ACCOA	acetyl-CoA	GLYCER	glycerate
AKG	α-ketoglutarate	GLYCO	glycolate
C16	palmitic acid	GLYOX	glyoxylate
C18	stearic acid	ICIT	isocitrate
C20	arachidic acid	INO	inositol
CIT	citrate	MAL	malate
CO2	internal CO_2	MTHF	5.10-methylenetetrahydrofolate
CO2EX	assimilated CO_2	OAA	oxaloacetate
CO2in	CO_2 in flux chamber	P5P	pentose 5-phosphate
CO2sink	respired CO_2	PEP	phosphoenolpyruvate

DHAP	dihydroxyacetone phosphate	PYR	pyruvate
E4P	erythrose 4-phosphate	RBP	ribulose 1.5-bisphosphate
F6P	fructose 6-phosphate	S7P	sedoheptulose 7-phosphate
FBP	fructose 1.6-bisphosphate	SBP	sedoheptulose 1.7-bisphosphate
FRC	fructose	SUCC	succinate
FUM	fumarate		

10.6. Translocation of assimilated carbon to the root (*O. sativa*)

To decipher which metabolites are responsible for the translocation of carbon to the root in rice

seedlings, GC-irMS analysis of root tissue was performed during a $^{13}CO_2$ labeling study. Several

amino acids and free sugars were analyzed, of which the results are shown in Fig. 10-1.

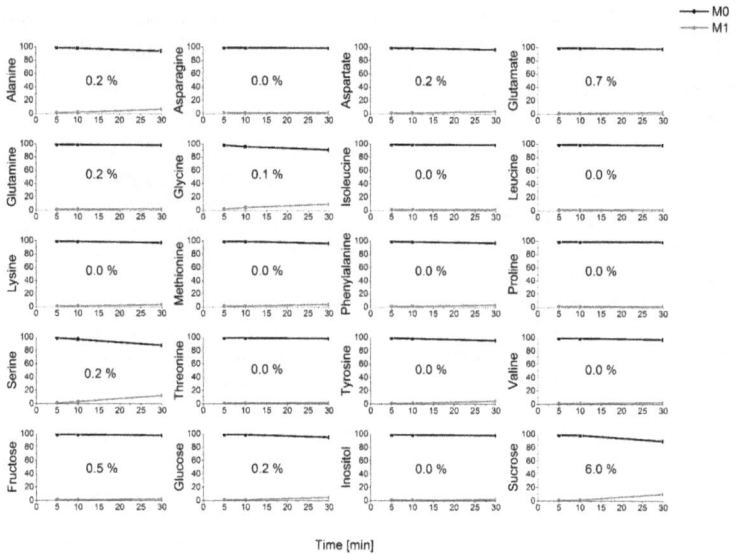

Figure 10-1: Measured transient mass isotopomer distribution of different sugars and amino acids in *O. sativa* roots as determined by GC-irMS

The mass isotopomer abundance is displayed in % over time (min). Additionally, the total carbon percentage translocating into the respective pool sizes are shown in the right upper corner of the respective graph.

10.7. Experimental biomass yield of *A. thaliana* from literature

Experimental values for the growth yield of *A. thaliana* were taken from Sulpice et al. (2014) and calculated as indicated in the following table.

Table 10-11: Supplemental data to Sulpice et al. (2014) with additional column for the biomass yield, which is calculated by the following formula: $\text{BMY} = \frac{\text{RGR} \cdot \text{DryMass}}{\text{net C assimilaton}}$.

photo-period	RGR	DRY MASS	net C assimilation per 24 h cycle	Calculated Biomass Yield daytime
h of light	*g FW /g FW /d*	*% FW*	*μmol C / g FW / d*	*g DW / mol C*
6	0.11351	8.94	420	24.2
8	0.17081	8.74	641	23.3
12	0.26000	9.14	944	25.2
18	0.3065	8.95	972	28.2

10.8. Biotechnological potential of *A. thaliana*

The biotechnological potential of *A. thaliana* leaves was assessed by evaluating the number of

calculated flux modes, theoretical maximum yield and carbon efficiency for each of the ten

valuable products (Table 10-12). In addition, genetic targets were predicted by Flux Design

(Table 10-13).

Table 10-12: Maximal theoretical product yield of 10 biotechnologically interesting products in autotrophic Arabidopsis leaves as predicted by the metabolic model.

	Cysteine	Iso-pentenyl pyrophosphate	Lignocellulose	Lysine	Methionine
Product Yield [g / C-mol Substrate]	40.35	27.32	33.26	24.37	37.30
Carbon Efficiency [%]	100	55	94	100	100
Number of EFMs	5 574 225	5 540 919	7 014 704	6 723 386	12 700 535

	Proline	Sucrose	Starch	Threonine	Tryptophan
Product Yield [g / C-mol Substrate]	76.79	13.50	28.51	29.78	11.35
Carbon Efficiency [%]	83	100	100	100	94
Number of EFMs	5 804 589	6 038 767	5 883 222	13 957 803	5 714 408

Table 10-13 (next page): Targets for product yield enhancement as predicted by Flux Design for defined subsets of modes. The mode subsets are visualized for better understanding by plotting the biomass synthesis against the product synthesis and color-coding the excluded modes red and the accepted flux distributions green. The plus or minus sign preceding the reaction number indicate a positive, resp. negative target potential coefficient. The enzyme names corresponding to the reaction numbers are listed below.

	all modes	product formation	no biomass but product forming modes	biomass and product forming modes	5 % range of maximum carbon efficiency	1 % range of maximum carbon efficiency
Cysteine	+3 +71	+1 +3 -4 +71	+3 -5 +71	-1 +3 -4 -20 +71	-1 +3 -4 -20 +65 +68 +71	-1 +3 -4 -5 +14 +18 +19 -20 -47 +65 +68 +71
Iso-pentenyl pyrophosphate	3	-1 +3 -4 +20 +71	+3 -5 +20 +71	+1 +3 +4 +20 +71	-1 +3 -4 +5 +20 +71	-1 +3 -4 +5 +18 +20 -47 +71
Ligno-cellulose	+3 +71	+1 +3 +4 +71	+3 -5 -47 +71	-1 +3 -4 -18 -20 +71	-1 +3 -4 -18 -20 +71	-1 +3 -4 -5 -14 -15 -18 -19 -20 -47 +71
Lysine	+3 +71	-1 +3 -4 +71	+3 -5 +71	+1 +3 +4 +71	-1 +3 -4 -20 +71	-1 +3 -4 -5 +14 +18 +19 -20 -47 +71
Methionine	+3 +71	-1 +3 -4 +71	+3 +71	-1 +3 -4 +71	-1 +3 -4 -20 +71	-1 +3 -4 -20 +71

Proline	+3 +71	+3 +71	+3 -5 +71	+1 +3 +4 +20 +71	-1 +3 -4 -5 -20 +47 +71	-1 -3 -4 +5 +14 +15 +18 +19 -20 +47 +71
Sucrose	+3 +71	+1 +3 +4 +71	+3 -5 +47 +71	+1 +3 +4 +20 +71	-1 +3 -4 -20 +71	-1 +3 -4 -5 +14 +18 +19 -20 -47 +71
Starch	3	-1 +3 -4 +71	+3 +5 +71	-1 +3 -4 -20 +71	-1 +3 -4 -5 -20	-1 +3 -4 -5 +18 -20 -47
Threonine	+3 +71	+1 +3 +4 +71	+3 -5 +71	-1 +3 -4 +71	-1 +3 -4 +71	-1 +3 -4 -20 +71
Tryptophan	+3 +71	-1 +3 -4 +71	+3 +5 +71	+3 +71	-1 +3 -4 -14 -15 -18 -20	-1 +3 -4 +5 -14 -15 -18 +19 -20 -47 +71

1	*in silico* biomass transporter	19	transketolase
3	*in silico* product transporter	20	plastidic pyruvate dehydrogenase
4	*in silico* biomass synthesis	47a	mitochondrial pyruvate dehydrogenase
5	CO_2 diffusion	47b	citrate synthase
14	plastidic ribose 5-phosphate isomerase	65	photosynthesis light reactions
15	plastidic ribulose-phosphate 3-epimerase	68	*in silico* photon transporter
18	transketolase	71	product synthesis

10.9. Results of ^{13}C-INST-MFA flux estimation and parameter continuation

In vivo fluxes were calculated for fifteen-day old wild type rice seedlings, as well as DMSO and Imazapyr-treated plants. In addition, flux estimation was performed for several scenarios, including different network topologies and levels of data coverage.

10.9.1. Flux results in wild type seedlings

Table 10-14: Fluxes in wild type seedlings determined by ^{13}C-INST-MFA. The flux values are expressed in mmol (100 mmol CO_2)$^{-1}$. The boundaries of the 95 % confidence intervals and the thereof derived standard deviation were determined through parameter continuation.

Reaction number	Reaction	WT06			
		val	lb	ub	std
'R1'	CO2in.p -> CO2EX.p	91.52	91.41	91.85	0.11
'R2'	12CO2in.p -> CO2in.p	5.90	5.41	6.21	0.20
'R3'	13CO2in.p -> CO2in.p	85.56	85.19	86.02	0.21
'R4'	CO2.p -> CO2EX.p	8.48	8.15	8.59	0.11
'R5'	CO2.p -> CO2sink.s	0.19	0.00	0.43	0.11
'R6'	biomass formation	2.68	2.68	2.68	0.00
'R7 net'	G6P.cp -> F6P.cp	-3.49	-3.60	-3.45	0.04
'R7 exch'	F6P.cp -> G6P.cp	NaN	127.10	Inf	FALSE
'R8 net'	FBP.cp -> F6P.cp	4.08	4.04	4.18	0.04
'R8 exch'	F6P.cp -> FBP.cp	3.89	3.53	5.01	0.38
'R9 net'	FBP.cp -> DHAP.cp + GAP.cp	-4.08	-4.18	-4.04	0.04
'R9 exch'	DHAP.cp + GAP.cp -> FBP.cp	0.56	0.00	1.22	0.31
'R10 net'	GAP.cp -> DHAP.cp	73.00	72.86	73.42	0.14
'R10 exch'	DHAP.cp -> GAP.cp	NaN	0.00	Inf	FALSE
'R11 net'	GAP.cp -> 3PG.cp	-182.16	-183.21	-181.85	0.35
'R11 exch'	3PG.cp -> GAP.cp	427.30	320.11	1013.80	176.96
'R12 net'	3PG.cp -> 2PG.cp	18.02	17.81	18.11	0.08
'R12 exch'	2PG.cp -> 3PG.cp	21.08	16.90	24.22	1.87
'R13 net'	2PG.cp -> PEP.cp	18.02	17.81	18.11	0.08
'R13 exch'	PEP.cp -> 2PG.cp	NaN	1251.27	Inf	FALSE
'R14 net'	PYR.cp -> PEP.cp	-15.42	-19.52	-7.34	3.11
'R14 exch'	PEP.cp -> PYR.cp	5.21	0.00	10.87	2.77
'R15'	PYR.cp -> ACCOA.p + CO2.p	2.17	2.17	2.17	0.00
'R16'	G6P.cp -> 6PG.cp	0.05	0.00	0.15	0.04
'R17'	6PG.cp -> P5P.p + CO2.p	0.05	0.00	0.15	0.04
'R18 net'	GAP.cp + S7P.p -> E4P.p + F6P.p	35.90	35.80	36.10	0.08
'R18 exch'	E4P.p + F6P.p -> GAP.cp + S7P.p	NaN	30.55	Inf	FALSE

'R19 net'	E4P.p + P5P.p -> F6P.p + GAP.cp	-35.90	-36.10	-35.80	0.08
'R19 exch'	F6P.p + GAP.cp -> E4P.p + P5P.p	NaN	112.00	Inf	FALSE
'R20 net'	GAP.cp + S7P.p -> P5P.p + P5P.p	33.06	32.97	33.27	0.08
'R20 exch'	P5P.p + P5P.p -> GAP.cp + S7P.p	0.57	0.00	1.02	0.26
'R21'	P5P.p -> RBP.p	100.70	100.50	101.30	0.20
'R22'	CO2EX.p + RBP.p -> 3PG.cp + 3PG.cp	100.00	100.00	100.00	0.00
'R23'	DHAP.cp + E4P.p -> S7P.p	68.96	68.77	69.37	0.15
'R24'	RBP.p -> GLYCO.pg + 3PG.cp	0.70	0.50	1.30	0.20
'R25'	GLYCO.pg -> GLYOX.g	0.70	0.50	1.30	0.20
'R26'	GLYOX.g + serine.p -> glycine.p + GLYCER.pg	0.13	0.00	0.86	0.22
'R27'	GLYCER.pg -> 3PG.cp	0.13	0.00	0.86	0.22
'R28'	sucrose.r -> zzzzzzzzzzzzzzzSink	0.47	0.46	0.49	0.01
R29 net	OAA.c -> CO2.p + PEP.cp	-0.10	-7.08	5.10	3.11
R29 exch	OAA.c <- CO2.p + PEP.cp	22.84	13.83	28.09	3.64
'R30 net'	MAL.p -> PYR.cp + CO2.p	-25.42	-45.25	0.88	11.77
'R30 exch'	PYR.cp + CO2.p -> MAL.p	21.38	11.05	33.84	5.81
'R31 net'	MAL.m -> MAL.c	-19.92	-39.56	-1.83	9.62
'R31 exch'	MAL.c -> MAL.m	NaN	0.00	Inf	FALSE
'R32 net'	PYR.m -> PYR.cp	7.22	-2.71	34.63	9.52
'R32 exch'	PYR.cp -> PYR.m	NaN	0.00	Inf	FALSE
'R33 net'	OAA.m + CIT.c -> OAA.c + CIT.m	0.00	0.00	0.00	0.00
'R33 exch'	OAA.c + CIT.m -> OAA.m + CIT.c	0.60	0.00	1.89	0.48
'R34 net'	MAL.m + OAA.c -> MAL.c + OAA.m	-2.49	-9.48	2.71	3.11
'R34 exch'	MAL.c + OAA.m -> MAL.m + OAA.c	10.12	5.02	18.82	3.52
'R35 net'	MAL.c -> MAL.p	-25.42	-45.25	0.88	11.77
'R35 exch'	MAL.p -> MAL.c	NaN	0.00	NaN	FALSE
'R36'	PYR.m -> ACCOA.m + CO2.p	2.62	2.40	2.71	0.08
'R37'	OAA.m + ACCOA.m -> CIT.m	2.62	2.40	2.71	0.08
'R38 net'	CIT.m -> AKG.m + CO2.p	2.60	2.39	2.69	0.08
'R38 exch'	AKG.m + CO2.p -> CIT.m	NaN	0.00	Inf	FALSE
'R39'	AKG.m -> SUCC.m + CO2.p	0.50	0.29	0.59	0.08
'R40'	SUCC.m -> FUM.m	0.50	0.28	0.59	0.08
'R41'	FUM.m -> MAL.m	0.88	0.66	0.97	0.08
'R42'	MAL.m -> OAA.m	8.08	0.00	12.19	3.11
'R43'	MAL.m -> CO2.p + PYR.m	18.09	0.00	37.73	9.62
'R44'	G6P.cp -> starch.p	0.10	0.10	0.10	0.00
'R45'	G6P.cp -> INO.cp	0.00	0.00	0.00	0.00
'R46'	G6P.cp -> GLC.cp	6.85	6.31	7.42	0.28
'R47'	GLC.cp -> G6P.cp	4.96	4.42	5.54	0.28
'R48'	F6P.cp -> FRC.cp	0.06	0.06	0.06	0.00
'R49'	G6P.cp + F6P.cp -> sucrose.cp	0.48	0.47	0.50	0.01

'R50'	sucrose.cp -> sucrose.r	0.47	0.46	0.49	0.01
'R51'	8*ACCOA.p -> C16.p	0.03	0.03	0.03	0.00
'R52'	9*ACCOA.p -> C18.p	0.15	0.15	0.15	0.00
'R53'	10*ACCOA.p -> C20.p	0.00	0.00	0.00	0.00
'R54'	3PG.cp -> serine.p	0.10	0.03	0.33	0.08
'R55'	3PG.cp -> cysteine.p	0.05	0.05	0.05	0.00
'R56'	3PG.cp -> glycine.p + CO2.p	0.32	0.22	0.61	0.10
'R57'	GLYOX.g -> glycine.p	0.51	0.44	0.54	0.03
'R58'	glycine.p -> zzMTHF.p + CO2.p	0.21	0.07	0.65	0.15
'R59'	glycine.p + zzMTHF.p -> serine.p	0.21	0.07	0.65	0.15
'R60'	PYR.cp + glutamate.m -> alanine.c + AKG.m	0.58	0.58	0.58	0.00
'R61'	PYR.cp + PYR.cp + glutamate.m -> valine.p + AKG.m + CO2.p	0.41	0.41	0.41	0.00
'R62'	PYR.cp + PYR.cp + ACCOA.p + glutamate.m -> leucine.p + CO2.p + CO2.p + AKG.m	0.55	0.55	0.55	0.00
'R63'	OAA.c + glutamate.m -> aspar.c + AKG.m	0.24	0.00	0.62	0.16
'R64'	OAA.c -> aspar.c	0.39	0.00	0.62	0.16
'R65'	P5P.p + CO2.p + glutamate.m -> histidine.p + AKG.m	0.13	0.13	0.13	0.00
'R66'	AKG.m -> proline.c	0.34	0.34	0.34	0.00
'R67'	AKG.m + glutamate.m + OAA.c + CO2.p -> arginine.c + AKG.m + FUM.m	0.38	0.38	0.38	0.00
'R68'	OAA.c + PYR.cp + glutamate.m -> lysine.p + CO2.p + AKG.m	0.33	0.33	0.33	0.00
'R69'	OAA.c -> threonine.p	0.35	0.35	0.35	0.00
'R70'	OAA.c + PYR.cp + glutamate.m -> isoleucine.p + CO2.p + AKG.m	0.27	0.27	0.27	0.00
'R71'	OAA.c + 3PG.cp + CO2.p -> methionine.p + PYR.cp	0.09	0.09	0.09	0.00
'R72'	E4P.p + PEP.cp + PEP.cp + glutamate.m -> tyrosine.p + CO2.p + AKG.m	0.77	0.77	0.77	0.00
'R73'	E4P.p + PEP.cp + PEP.cp + glutamate.m -> phenylalanine.p + CO2.p + AKG.m	2.06	2.06	2.06	0.00
'R74'	AKG.m -> glutamate.m	6.58	6.34	6.97	0.16
'R75'	E4P.p + PEP.cp + PEP.cp + P5P.p + 3PG.cp + glutamate.m -> CO2.p + tryptophane.p + PYR.cp + GAP.cp + AKG.m	0.00	0.00	0.00	0.00

10.9.2. Flux results for wild type seedlings, 4 hours after DMSO or Imazapyr-treatment

Table 10-15 (next page): Fluxes in wild type seedlings after treatment with DMSO or Imazapyr, rescpectively. All fluxes were determined by ^{13}C-INST-MFA and are expressed in mmol (100 mmol CO_2)$^{-1}$. The boundaries of the 95 % confidence intervals and the thereof derived standard deviation were determined through parameter continuation.

Reaction number	Reaction	DMSO				IMAZAPYR			
		val	lb	ub	std	val	lb	ub	std
'R1	'CO2in.p -> CO2EX.p	93.13	91.82	95.37	0.90	92.78	92.62	93.28	0.17
'R2	'2CO2in.p -> CO2in.p	5.02	3.99	5.73	0.44	5.11	4.58	5.53	0.24
'R3	'3CO2in.p -> CO2in.p	88.33	86.41	90.20	0.97	87.66	87.11	88.25	0.29
'R4	'CO2.p -> CO2EX.p	6.87	4.63	8.18	0.90	7.22	6.72	7.38	0.17
'R5	'CO2.p -> CO2sink.s	2.31	0.00	3.55	0.91	0.29	0.00	0.65	0.17
'R6	'biomass equation	2.68	2.68	2.68	0.00	2.68	2.68	2.68	0.00
'R7 net	'G6P.cp -> F6P.cp	-3.64	-4.32	-3.38	0.24	-3.60	-3.69	-3.50	0.05
'R8 net	'FBP.cp -> F6P.cp	4.05	3.93	4.10	0.04	4.21	4.11	4.30	0.05
'R9 net	'FBP.cp -> DHAP.cp + GAP.cp	-4.05	-4.10	-3.93	0.04	-4.21	-4.30	-4.11	0.05
'R10 net	'GAP.cp -> DHAP.cp	73.26	72.92	73.64	0.18	73.21	73.13	73.34	0.05
'R11 net	'GAP.cp -> 3PG.cp	-182.68	-183.82	-182.04	0.45	-182.67	-182.97	-182.45	0.13
'R12 net	'3PG.cp -> 2PG.cp	17.89	17.62	17.95	0.09	17.60	17.55	17.62	0.02
'R13 net	'2PG.cp -> PEP.cp	17.89	17.62	17.95	0.09	17.60	17.55	17.62	0.02
'R14 net	'PYR.cp -> PEP.cp	-8.13	-9.64	-7.54	0.54	-8.04	-8.88	-7.55	0.34
'R15	'PYR.cp -> ACCOA.p + CO2.p	2.23	2.23	2.23	0.00	2.29	2.29	2.29	0.00
'R16	'G6P.cp -> 6PG.cp	0.45	0.00	0.94	0.24	0.08	0.00	0.22	0.06
'R17	'6PG.cp -> P5P.p + CO2.p	0.45	0.00	0.94	0.24	0.08	0.00	0.22	0.06
'R18 net	'GAP.cp + S7P.p -> E4P.p + F6P.p	35.98	35.63	36.26	0.16	35.91	35.84	36.00	0.04
'R19 net	'E4P.p + P5P.p -> F6P.p + GAP.cp	-35.98	-36.26	-35.63	0.16	-35.91	-36.00	-35.84	0.04
'R20 net	'GAP.cp + S7P.p -> P5P.p + P5P.p	33.16	32.82	33.45	0.16	33.14	33.07	33.23	0.04
'R21	'P5P.p -> RBP.p	101.09	100.76	101.84	0.28	100.98	100.83	101.15	0.08
'R22	'CO2EX.p + RBP.p -> 3PG.cp + 3PG.cp	100.00	100.00	100.00	0.00	100.00	100.00	100.00	0.00
'R23	'DHAP.cp + E4P.p -> S7P.p	69.14	68.45	69.71	0.32	69.05	68.91	69.23	0.08
'R24	'RBP.p -> GLYCO.pg + 3PG.cp	1.09	0.76	1.84	0.28	0.98	0.83	1.15	0.08
'R25	'GLYCO.pg -> GLYOX.g	1.09	0.76	1.84	0.28	0.98	0.83	1.15	0.08
'R26	'GLYOX.g + serine.p -> glycine.p + GLYCER.pg	0.54	0.00	1.01	0.26	0.06	0.00	0.22	0.06
'R27	'GLYCER.pg -> 3PG.cp	0.54	0.00	1.01	0.26	0.06	0.00	0.22	0.06
'R28	'sucrose.r -> zzzzzzzzzzzzSink	0.43	0.43	0.51	0.02	0.50	0.50	0.51	0.01

Reaction								
R29 net'OAA.c -> CO2.p + PEP.cp	-4.22	-4.80	-2.28	0.64	-4.29	-4.55	-3.93	0.16
'R30 net'MAL.p -> PYR.cp + CO2.p	0.45	0.01	0.54	0.13	0.14	-0.13	0.29	0.11
'R31 net'MAL.m -> MAL.c	-2.92	-4.00	-1.85	0.55	-2.13	-2.53	-1.89	0.16
'R32 net'PYR.m -> PYR.cp	-1.98	-2.40	-0.43	0.50	-2.16	-2.17	-2.15	0.01
'R33 net'OAA.m + CIT.c -> OAA.c + CIT.m	0.00	0.00	0.00	0.00	0.00	0.00	0.00	0.00
'R34 net'MAL.m + OAA.c -> MAL.c + OAA.m	1.85	0.02	2.39	2.36	1.91	1.56	2.17	0.16
'R35 net'MAL.c -> MAL.p	0.45	0.01	2.39	0.61	0.14	-0.13	0.29	0.11
'R36'PYR.m -> ACCOA.m + CO2.p	2.33	2.05	2.39	0.09	2.15	2.10	2.17	0.02
'R37'OAA.m + ACCOA.m -> CIT.m	2.33	2.05	2.39	0.09	2.15	2.10	2.17	0.02
'R38 net'CIT.m -> AKG.m + CO2.p	2.31	2.03	2.37	0.09	2.13	2.08	2.16	0.02
'R39'AKG.m -> SUCC.m + CO2.p	0.28	0.00	0.34	0.09	0.06	0.00	0.08	0.02
'R40'SUCC.m -> FUM.m	0.27	0.00	0.34	0.09	0.06	0.00	0.08	0.02
'R41'FUM.m -> MAL.m	0.58	0.30	0.64	0.09	0.36	0.30	0.38	0.02
'R42'MAL.m -> OAA.m	0.59	0.00	2.11	0.54	0.65	0.00	1.89	0.48
'R43'MAL.m -> CO2.p + PYR.m	1.07	0.00	2.16	0.55	0.24	0.00	0.64	0.16
'R44'G6P.cp -> starch.p	0.07	0.07	0.07	0.00	0.12	0.12	0.12	0.00
'R45'G6P.cp -> INO.cp	0.00	0.00	0.00	0.00	0.00	0.00	0.00	0.00
'R46'G6P.cp -> GLC.cp	6.65	5.47	8.50	0.77	8.41	5.57	12.05	1.65
'R47'GLC.cp -> G6P.cp	4.76	3.58	6.61	0.77	6.52	3.68	10.16	1.65
'R48'F6P.cp -> FRC.cp	0.05	0.05	0.05	0.00	0.05	0.05	0.05	0.00
'R49'G6P.cp + F6P.cp -> sucrose.cp	0.49	0.43	0.51	0.02	0.51	0.50	0.51	0.01
'R50'sucrose.cp -> sucrose.r	0.43	0.43	0.51	0.02	0.50	0.50	0.51	0.00
'R51'8*ACCOA.p -> C16.p	0.03	0.03	0.03	0.00	0.03	0.03	0.03	0.00
'R52'9*ACCOA.p -> C18.p	0.16	0.16	0.16	0.00	0.16	0.16	0.16	0.00
'R53'10*ACCOA.p -> C20.p	0.00	0.00	0.00	0.00	0.00	0.00	0.00	0.03
'R54'3PG.cp -> serine.p	0.31	0.10	0.51	0.10	0.13	0.09	0.19	0.00
'R55'3PG.cp -> cysteine.p	0.04	0.04	0.04	0.00	0.10	0.10	0.10	0.01
'R56'3PG.cp -> glycine.p + CO2.p	0.56	0.32	0.83	0.13	0.33	0.29	0.34	0.03
'R57'GLYOX.g -> glycine.p	0.76	0.65	0.91	0.07	0.88	0.83	0.94	0.03
'R58'glycine.p -> zzMTHF.p + CO2.p	0.54	0.26	0.94	0.17	0.33	0.28	0.44	0.04
'R59'glycine.p + zzMTHF.p -> serine.p	0.54	0.26	0.94	0.17	0.33	0.28	0.44	0.04

ID	Reaction								
R60	PYR.cp + glutamate.m -> alanine.c + AKG.m	0.61	0.61	0.61	0.00	0.61	0.61	0.61	0.00
R61	PYR.cp + PYR.cp + glutamate.m -> valine.p + AKG.m + CO2.p	0.43	0.43	0.43	0.00	0.41	0.41	0.41	0.00
R62	PYR.cp + PYR.cp + ACCOA.p + glutamate.m -> leucine.p + CO2.p + CO2.p + AKG.m	0.56	0.56	0.56	0.00	0.55	0.55	0.55	0.00
R63	OAA.c + glutamate.m -> aspar.c + AKG.m	0.45	0.00	0.67	0.17	0.15	0.00	0.67	0.17
R64	OAA.c -> aspar.c	0.22	0.00	0.67	0.17	0.52	0.00	0.67	0.17
R65	P5P.p + CO2.p + glutamate.m -> histidine.p + AKG.m	0.13	0.13	0.13	0.00	0.13	0.13	0.13	0.00
R66	AKG.m -> proline.c	0.35	0.35	0.35	0.00	0.34	0.34	0.34	0.00
R67	AKG.m + glutamate.m + OAA.c + CO2.p -> arginine.c + AKG.m + FUM.m	0.30	0.30	0.30	0.00	0.30	0.30	0.30	0.00
R68	OAA.c + PYR.cp + glutamate.m -> lysine.p + CO2.p + AKG.m	0.35	0.35	0.35	0.00	0.34	0.34	0.34	0.00
R69	OAA.c -> threonine.p	0.35	0.35	0.35	0.00	0.35	0.35	0.35	0.00
R70	OAA.c + PYR.cp + glutamate.m -> isoleucine.p + CO2.p + AKG.m	0.28	0.28	0.28	0.00	0.28	0.28	0.28	0.00
R71	OAA.c + 3PG.cp + CO2.p -> methionine.p + PYR.cp	0.09	0.09	0.09	0.00	0.09	0.09	0.09	0.00
R72	E4P.p + PEP.cp + PEP.cp + glutamate.m -> tyrosine.p + CO2.p + AKG.m	0.78	0.78	0.78	0.00	0.73	0.73	0.73	0.00
R73	E4P.p + PEP.cp + PEP.cp + glutamate.m -> phenylalanine.p + CO2.p + AKG.m	2.04	2.04	2.04	0.00	2.03	2.03	2.03	0.00
R74	AKG.m -> glutamate.m	6.77	6.33	7.00	0.17	6.43	6.28	6.95	0.17
R75	E4P.p + PEP.cp + PEP.cp + P5P.p + 3PG.cp + glutamate.m -> CO2.p + tryptophane.p + PYR.cp + GAP.cp + AKG.m	0.00	0.00	0.00	0.00	0.00	0.00	0.00	0.00

10.9.3. Investigation of different network topologies

Table 10-16 (next page): Fluxes calculated with different metabolic network topologies. No hexokinase: network without hexokinase function. F6P.cp: no compartmentation of F6P. Parallel: Parallel network including a hexokinase function. Parallel no Hexokinase: Parallel network model excluding a hexokinase function. All fluxes were determined by ^{13}C-INST-MFA and are expressed in mmol (100 mmol CO_2)$^{-1}$. The boundaries of the 95 % confidence intervals and the thereof derived standard deviation were determined through parameter continuation.

Reaction number	Reaction	no Hexokinase				F6P.cp			
		val	lb	ub	std	val	lb	ub	std
'R1	CO2in.p -> CO2EX.p	91.59	91.41	92.16	0.19	94.78	91.53	100.00	2.16
'R2	12CO2in.p -> CO2in.p	5.80	5.23	6.30	0.27	5.65	5.01	6.32	0.33
'R3	13CO2in.p -> CO2in.p	85.65	85.12	86.30	0.30	90.81	86.09	95.15	2.31
'R4	CO2.p -> CO2EX.p	8.41	7.84	8.59	0.19	5.22	0.00	8.47	2.16
'R5	CO2.p -> CO2sink.s	0.38	0.00	0.78	0.20	4.00	0.00	8.59	2.19
'R6	biomass equation	2.68	2.68	2.68	0.00	2.68	2.68	2.68	0.00
'R7 net	G6P.cp -> F6P.cp	-3.50	-3.67	-3.45	0.06	-3.52	-3.61	-3.46	0.04
'R8 net	FBP.cp -> F6P.cp	4.08	4.04	4.21	0.04	-0.27	-2.06	1.90	1.01
'R9 net	FBP.cp -> DHAP.cp + GAP.cp	-4.08	-4.21	-4.04	0.04	0.27	-1.90	2.06	1.01
'R10 net	GAP.cp -> DHAP.cp	73.03	72.86	73.65	0.20	72.93	72.86	73.11	0.06
'R11 net	GAP.cp -> 3PG.cp	-182.56	-183.72	-181.84	0.48	-182.03	-182.45	-181.83	0.16
'R12 net	3PG.cp -> 2PG.cp	18.03	17.70	18.12	0.11	17.99	17.62	18.12	0.13
'R13 net	2PG.cp -> PEP.cp	18.03	17.70	18.12	0.11	17.99	17.62	18.12	0.13
'R14 net	PYR.cp -> PEP.cp	-17.37	-19.96	-7.34	3.22	-11.04	-16.16	-7.34	2.25
'R15	PYR.cp -> ACCOA.p + CO2.p	2.17	2.17	2.17	0.00	2.17	2.17	2.17	0.00
'R16	G6P.cp -> 6PG.cp	0.07	0.00	0.18	0.05	0.04	0.00	0.16	0.04
'R17	6PG.cp -> P5P.p + CO2.p	0.07	0.00	0.18	0.05	0.04	0.00	0.16	0.04
'R18 net	GAP.cp + S7P.p -> E4P.p + F6P.p	35.88	35.83	36.24	0.11	39.03	37.33	40.71	0.86
'R19 net	E4P.p + P5P.p -> F6P.p + GAP.cp	-35.88	-36.24	-35.83	0.11	-35.84	-35.95	-35.79	0.04
'R20 net	GAP.cp + S7P.p -> P5P.p + P5P.p	33.05	32.99	33.41	0.11	33.02	32.95	33.12	0.04
'R21	P5P.p -> RBP.p	100.74	100.49	101.56	0.27	100.59	100.48	100.86	0.10
'R22	CO2EX.p + RBP.p -> 3PG.cp + 3PG.cp	100.00	100.00	100.00	0.00	100.00	100.00	100.00	0.00
'R23	DHAP.cp + E4P.p -> S7P.p	68.93	68.82	69.65	0.21	73.05	70.26	74.56	1.10
'R24	RBP.p -> GLYCO.pg + 3PG.cp	0.74	0.49	1.56	0.27	0.59	0.48	0.86	0.10
'R25	GLYCO.pg -> GLYOX.g	0.74	0.49	1.56	0.27	0.59	0.48	0.86	0.10
'R26	GLYOX.g + serine.p -> glycine.p + GLYCER.pg	0.20	0.00	1.05	0.27	0.07	0.00	0.36	0.09
'R27	GLYCER.pg -> 3PG.cp	0.20	0.00	1.05	0.27	0.07	0.00	0.36	0.09
'R28	sucrose.r -> zzzzzzzzzzzzSink	0.47	0.47	0.52	0.01	0.47	0.47	0.51	0.01

Reaction								
R29 net OAA.c -> CO2.p + PEP.cp	0.79	-5.11	7.50	3.22	-4.03	-5.11	2.80	2.02
R30 net MAL.p -> PYR.cp + CO2.p	-17.43	-45.80	0.88	11.91	-4.25	-21.01	0.61	5.52
R31 net MAL.m -> MAL.c	-23.43	-41.59	-1.83	10.14	-7.20	-22.40	-2.09	5.18
R32 net PYR.m -> PYR.cp	12.86	-2.71	36.53	10.01	2.95	-2.43	13.53	4.07
R33 net OAA.m + CIT.c -> OAA.c + CIT.m	0.00	0.00	0.00	0.00	0.00	0.00	0.00	0.00
R34 net MAL.m + OAA.c -> MAL.c + OAA.m	-3.18	-9.89	2.71	3.22	1.64	-5.19	2.72	2.02
R35 net MAL.c -> MAL.p	-17.43	-45.80	0.88	11.91	-4.25	-21.01	0.61	5.52
'R36'PYR.m -> ACCOA.m + CO2.p	2.62	2.29	2.71	0.11	2.58	2.22	2.72	0.13
'R37'OAA.m + ACCOA.m -> CIT.m	2.62	2.29	2.71	0.11	2.58	2.22	2.72	0.13
R38 net CIT.m -> AKG.m + CO2.p	2.61	2.28	2.70	0.11	2.57	2.20	2.70	0.13
'R39'AKG.m -> SUCC.m + CO2.p	0.51	0.18	0.60	0.11	0.47	0.10	0.60	0.13
'R40'SUCC.m -> FUM.m	0.50	0.18	0.60	0.11	0.46	0.10	0.60	0.13
'R41'FUM.m -> MAL.m	0.88	0.55	0.98	0.11	0.84	0.48	0.98	0.13
'R42'MAL.m -> OAA.m	10.03	0.00	12.62	3.22	3.70	0.00	8.83	2.25
'R43'MAL.m -> CO2.p + PYR.m	21.60	0.00	39.76	10.14	5.37	0.25	20.56	5.18
'R44'G6P.cp -> starch.p	0.10	0.10	0.10	0.00	0.10	0.10	0.10	0.00
'R45'G6P.cp -> INO.cp	0.00	0.00	0.00	0.00	0.00	0.00	0.00	0.00
'R46'G6P.cp -> GLC.cp	1.89	1.89	1.89	0.00	10.62	9.67	11.72	0.52
'R47'GLC.cp -> G6P.cp	0.00	0.00	0.00	0.00	8.73	7.78	9.83	0.52
'R48'F6P.cp -> FRC.cp	0.06	0.06	0.06	0.00	0.06	0.06	0.06	0.00
'R49'G6P.cp + F6P.cp -> sucrose.cp	0.48	0.47	0.52	0.01	0.48	0.47	0.51	0.01
'R50'sucrose.cp -> sucrose.r	0.47	0.47	0.52	0.01	0.47	0.47	0.51	0.01
'R51'8*ACCOA.p -> C16.p	0.03	0.03	0.03	0.00	0.03	0.03	0.03	0.00
'R52'9*ACCOA.p -> C18.p	0.15	0.15	0.15	0.00	0.15	0.15	0.15	0.00
'R53'10*ACCOA.p -> C20.p	0.00	0.00	0.00	0.00	0.00	0.00	0.00	0.00
'R54'3PG.cp -> serine.p	0.09	0.03	0.42	0.10	0.07	0.03	0.16	0.03
'R55'3PG.cp -> cysteine.p	0.05	0.05	0.05	0.00	0.05	0.05	0.05	0.00
'R56'3PG.cp -> glycine.p + CO2.p	0.36	0.21	0.80	0.15	0.30	0.20	0.40	0.05
'R57'GLYOX.g -> glycine.p	0.50	0.40	0.55	0.04	0.52	0.47	0.56	0.02
'R58'glycine.p -> zzMTHF.p + CO2.p	0.13	0.06	0.89	0.21	0.15	0.06	0.32	0.06
'R59'glycine.p + zzMTHF.p -> serine.p	0.13	0.06	0.89	0.21	0.15	0.06	0.32	0.06

	Reaction								
R60	PYR.cp + glutamate.m -> alanine.c + AKG.m	0.58	0.58	0.58	0.00	0.58	0.58	0.58	0.00
R61	PYR.cp + PYR.cp + glutamate.m -> valine.p + AKG.m + CO2.p	0.41	0.41	0.41	0.00	0.41	0.41	0.41	0.00
R62	PYR.cp + PYR.cp + ACCOA.p + glutamate.m -> leucine.p + CO2.p + CO2.p + AKG.m	0.55	0.55	0.55	0.00	0.55	0.55	0.55	0.00
R63	OAA.c + glutamate.m -> aspar.c + AKG.m	0.23	0.00	0.62	0.16	0.29	0.00	0.62	0.16
R64	OAA.c -> aspar.c	0.39	0.00	0.62	0.16	0.29	0.00	0.62	0.16
R65	P5P.p + CO2.p + glutamate.m -> histidine.p + AKG.m	0.13	0.13	0.13	0.00	0.13	0.13	0.13	0.00
R66	AKG.m -> proline.c	0.34	0.34	0.34	0.00	0.34	0.34	0.34	0.00
R67	AKG.m + glutamate.m + OAA.c + CO2.p -> arginine.c + AKG.m + FUM.m	0.38	0.38	0.38	0.00	0.38	0.38	0.38	0.00
R68	OAA.c + PYR.cp + glutamate.m -> lysine.p + CO2.p + AKG.m	0.33	0.33	0.33	0.00	0.33	0.33	0.33	0.00
R69	OAA.c -> threonine.p	0.35	0.35	0.35	0.00	0.35	0.35	0.35	0.00
R70	OAA.c + PYR.cp + glutamate.m -> isoleucine.p + CO2.p + AKG.m	0.27	0.27	0.27	0.00	0.27	0.27	0.27	0.00
R71	OAA.c + 3PG.cp + CO2.p -> methionine.p + PYR.cp	0.09	0.09	0.09	0.00	0.09	0.09	0.09	0.00
R72	E4P.p + PEP.cp + PEP.cp + glutamate.m -> tyrosine.p + CO2.p + AKG.m	0.77	0.77	0.77	0.00	0.77	0.77	0.77	0.00
R73	E4P.p + PEP.cp + PEP.cp + glutamate.m -> phenylalanine.p + CO2.p + AKG.m	2.06	2.06	2.06	0.00	2.06	2.06	2.06	0.00
R74	AKG.m -> glutamate.m	6.57	6.34	6.97	0.16	6.68	6.34	6.97	0.16
R75	E4P.p + PEP.cp + PEP.cp + P5P.p + 3PG.cp + glutamate.m -> CO2.p + tryptophane.p + PYR.cp + GAP.cp + AKG.m	0.00	0.00	0.00	0.00	0.00	0.00	0.00	0.00

Reaction number	Reaction	Parallel				Parallel no Hexokinase			
		val	lb	ub	std	val	lb	ub	std
R1	CO2in.p -->CO2EX.p	91.65	91.41	92.25	0.21	91.97	91.41	92.25	0.21
R2	12CO2in.p -->CO2in.p	6.04	5.25	6.73	0.38	5.31	5.25	6.73	0.38
R3	13CO2in.p -->CO2in.p	85.51	84.75	86.24	0.38	85.13	84.75	86.24	0.38
R4	CO2.p -->CO2EX.p	8.35	7.75	8.59	0.21	8.03	7.75	8.59	0.21
R5	CO2.p -->CO2sink.s	0.41	0.00	0.89	0.23	0.90	0.00	0.89	0.23
R6	biomass equation	2.68	2.68	2.68	0.00	2.68	2.68	2.68	0.00
R7 net	G6P.c -->F6P.c	-1.22	-7.10	7.63	3.76	NaN	-4.17	-3.35	0.21
R8 net	FBP.c -->F6P.c	1.50	-7.30	7.63	3.81	2.18	-1.22	5.28	1.66
R9 net	FBP.c -->DHAP.c + GAP.c	-1.50	-7.63	7.30	3.81	-2.18	-5.28	1.22	1.66
R10 net	GAP.c -->DHAP.c	4.86	-40.89	65.86	27.23	-0.12	-19.17	14.27	8.53
R11 net	GAP.c -->3PG.c	-2.32	-72.16	52.22	31.73	-4.92	-16.99	15.25	8.22
R12 net	3PG.c -->2PG.c	13.19	1.57	22.52	5.34	9.67	5.12	14.57	2.41
R13 net	2PG.c -->PEP.c	12.41	1.19	22.11	5.34	9.67	5.12	14.57	2.41
R14 net	PYR.c -->PEP.c	-8.72	-32.91	18.11	13.02	3.36	-1.38	9.48	2.77
R15	PYR.p -->ACCOA.p + CO2.p	2.17	2.17	2.17	0.00	2.17	2.17	2.17	0.00
R16	G6P.p -->6PG.p	0.74	0.00	1.85	0.47	1.64	0.63	1.72	0.28
R17	6PG.p -->P5P.p + CO2.p	0.74	0.00	1.85	0.47	1.64	0.63	1.72	0.28
R18 net	GAP.p + S7P.p -->E4P.p + F6P.p	NaN	-33.70	Inf	FALSE	12.98	4.90	31.59	6.81
R19 net	E4P.p + P5P.p -->F6P.p + GAP.p	-35.57	-36.52	-35.25	0.32	-35.30	-35.85	-35.25	0.15
R20 net	GAP.p + S7P.p -->P5P.p + P5P.p	32.63	32.40	33.71	0.33	32.47	32.42	33.01	0.15
R21	P5P.p -->RBP.p	100.94	100.40	102.60	0.56	100.69	100.50	101.40	0.23
R22	CO2EX.p + RBP.p -->3PG.p + 3PG.p	100.00	100.00	100.00	0.00	100.00	100.00	100.00	0.00
R23	DHAP.p + E4P.p -->S7P.p	69.82	0.00	229.00	58.42	44.34	0.00	61.10	15.59
R24	RBP.p -->GLYCO.pg + 3PG.p	0.95	0.33	2.64	0.59	0.69	0.47	1.40	0.24
R25	GLYCO.pg -->GLYOX.g	0.95	0.33	2.64	0.59	0.69	0.47	1.40	0.24
R26	GLYOX.g + serine.p -->glycine.p + GLYCER.pg	0.87	0.00	2.15	0.55	0.09	0.00	0.74	0.19
R27	GLYCER.pg -->3PG.p	0.87	0.00	2.15	0.55	0.09	0.00	0.74	0.19
R28	sucrose.r -->zzzzzzzzzzzzSink	0.47	0.47	0.51	0.01	0.47	0.47	0.50	0.01

Reaction								
R29 net 'OAA.c -->CO2.p + PEP.c	NaN	-5.07	8.93	3.57	-2.68	-4.86	2.96	1.99
R30 net 'MAL.p -->PYR.p + CO2.p	NaN	-76.04	0.85	19.61	-4.67	-10.06	0.49	2.69
R31 net 'MAL.m -->MAL.c	-13.64	-38.45	-1.83	9.34	-6.95	-10.97	-1.83	2.33
R32 net 'PYR.m -->PYR.c	5.48	-2.59	63.12	16.76	1.36	-2.24	9.16	2.91
R33 net 'OAA.m + CIT.c -->OAA.c + CIT.m	0.00	0.00	0.00	0.00	0.00	0.00	0.00	0.00
R34 net 'MAL.m + OAA.c -->MAL.c + OAA.m	1.54	-8.89	2.60	2.93	0.29	-5.35	2.47	1.99
R35 net 'MAL.c -->MAL.p	-6.95	-63.33	0.80	16.36	-4.67	-10.06	0.49	2.69
'R36 'PYR.m -->ACCOA.m + CO2.p	NaN	2.12	2.68	0.14	2.26	2.12	2.49	0.10
'R37 'OAA.m + ACCOA.m -->CIT.m	NaN	2.12	2.68	0.14	2.26	2.12	2.49	0.10
'R38 net 'CIT.m -->AKG.m + CO2.p	NaN	2.11	2.67	0.14	2.24	2.10	2.48	0.10
'R39 'AKG.m -->SUCC.m + CO2.p	NaN	0.00	0.59	0.15	0.14	0.00	0.38	0.10
'R40 'SUCC.m -->FUM.m	NaN	0.00	0.58	0.15	0.14	0.00	0.38	0.10
'R41 'FUM.m -->MAL.m	0.82	0.38	0.87	0.13	0.52	0.38	0.75	0.10
'R42 'MAL.m -->OAA.m	7.44	0.00	13.91	3.55	3.17	0.00	6.91	1.76
'R43 'MAL.m -->CO2.p + PYR.m	30.76	0.00	54.48	13.90	5.12	0.00	9.14	2.33
'R44 'G6P.p -->starch.p	0.10	0.10	0.10	0.00	0.10	0.10	0.10	0.00
'R45 'G6P.c -->INO.cp	0.00	0.00	0.00	0.00	0.00	0.00	0.00	0.00
'R46 'G6P.c -->GLC.cp	0.00	0.00	0.00	0.00	0.00	0.00	0.00	0.00
'R47 'GLC.cp -->G6P.c	NaN	1.89Inf	0.06	FALSE	1.89	1.89	1.89	0.00
'R48 'F6P.c -->FRC.cp	NaN	0.00Inf	0.51	FALSE	0.00	0.00	0.00	0.00
'R49 'G6P.c + F6P.c -->sucrose.cp	0.48	0.47	0.51	0.00	0.48	0.47	0.06	0.01
'R50 'sucrose.cp -->sucrose.r	0.47	0.47	0.03	0.01	0.47	0.47	0.50	0.01
'R51 '8*ACCOA.p -->C16.p	0.03	0.03	0.15	0.01	0.03	0.03	0.50	0.00
'R52 '9*ACCOA.p -->C18.p	0.15	0.15	0.00	0.00	0.15	0.15	0.03	0.00
'R53 '10*ACCOA.p -->C20.p	0.00	0.00	0.00	0.00	0.00	0.00	0.15	0.00
'R54 '3PG.p -->serine.p	0.22	0.02	1.00	0.25	0.08	0.03	0.32	0.08
'R55 '3PG.p -->cysteine.p	0.05	0.05	0.05	0.00	0.05	0.05	0.05	0.00
'R56 '3PG.p -->glycine.p + CO2.p	0.43	0.14	1.13	0.25	0.32	0.20	0.61	0.10
'R57 'GLYOX.g -->glycine.p	0.45	0.00	0.76	0.19	0.50	0.41	0.56	0.04
'R58 'glycine.p -->zzMTHF.p + CO2.p	0.55	0.05	1.36	0.33	0.16	0.06	0.60	0.14
'R59 'glycine.p + zzMTHF.p -->serine.p	0.55	0.05	1.36	0.33	0.16	0.06	0.60	0.14

	V1	V2	V3	V4	V5	V6	V7	V8	V9
'R60' PYR.c + glutamate.m -->alanine.c + AKG.m	0.58	0.58	0.58	0.58	0.00	0.58	0.58	0.58	0.58
'R61' PYR.p + PYR.p + glutamate.m -->valine.p + AKG.m + CO2.p	0.41	0.41	0.41	0.41	0.00	0.41	0.41	0.41	0.41
'R62' PYR.p + PYR.p + ACCOA.p + glutamate.m -->leucine.p + CO2.p + CO2.p + AKG.m	0.55	0.55	0.55	0.55	0.00	0.55	0.55	0.55	0.55
'R63' OAA.c + glutamate.m -->aspar.c + AKG.m	0.19	0.00	0.00	0.19	0.16	0.62	0.00	0.00	0.62
'R64' OAA.c -->aspar.c	0.44	0.00	0.00	0.44	0.16	0.62	0.00	0.00	0.62
'R65' P5P.p + CO2.p + glutamate.m -->histidine.p + AKG.m	0.13	0.13	0.13	0.13	0.00	0.13	0.13	0.13	0.13
'R66' AKG.m -->proline.c	0.34	0.34	0.34	0.34	0.00	0.34	0.34	0.34	0.34
'R67' AKG.m + glutamate.m + OAA.c + CO2.p -->arginine.c + AKG.m + FUM.m	0.38	0.38	0.38	0.38	0.00	0.38	0.38	0.38	0.38
'R68' OAA.c + PYR.p + glutamate.m -->lysine.p + CO2.p + AKG.m	0.33	0.33	0.33	0.33	0.00	0.33	0.33	0.33	0.33
'R69' OAA.c -->threonine.p	0.35	0.35	0.35	0.35	0.00	0.35	0.35	0.35	0.35
'R70' OAA.c + PYR.p + glutamate.m -->isoleucine.p + CO2.p + AKG.m	0.27	0.27	0.27	0.27	0.00	0.27	0.27	0.27	0.27
'R71' OAA.c + 3PG.p + CO2.p -->methionine.p + PYR.p	0.09	0.09	0.09	0.09	0.00	0.09	0.09	0.09	0.09
'R72' E4P.p + PEP.p + PEP.p + glutamate.m -->tyrosine.p + CO2.p + AKG.m	0.77	0.77	0.77	0.77	0.00	0.77	0.77	0.77	0.77
'R73' E4P.p + PEP.p + PEP.p + glutamate.m -->phenylalanine.p + CO2.p + AKG.m	2.06	2.06	2.06	2.06	0.00	2.06	2.06	2.06	2.06
'R74' AKG.m -->glutamate.m	6.53	6.34	6.53	6.53	0.16	6.97	6.34	6.53	6.97
'R75' E4P.p + PEP.p + PEP.p + P5P.p + 3PG.p + glutamate.m -->CO2.p + tryptophane.p + PYR.p + GAP.p + AKG.m	0.00	0.00	0.00	0.00	0.00	0.00	0.00	0.00	0.00

'R131 net'G6P.p -->F6P.p	-3.79	-11.43	3.35	3.77	-4.22	-6.84	-0.23	1.69
'R132 net'FBP.p -->F6P.p	40.07	-160.00	80.50	61.35	35.41	11.03	41.38	7.74
'R133 net'FBP.p -->DHAP.p + GAP.p	-40.07	80.50	-160.00	-61.35	-35.41	-41.38	-11.03	7.74
'R134 net'GAP.p -->DHAP.p	69.55	6.43	128.10	31.04	80.17	60.82	94.01	8.47
'R135 net'GAP.p -->3PG.p	-189.94	-233.90	-152.80	20.69	-173.79	-174.60	-165.10	2.42
'R136 net'3PG.p -->2PG.p	2.34	-18.82	15.72	8.81	6.23	3.80	8.42	1.18
'R137 net'2PG.p -->PEP.p	2.34	-18.82	15.72	8.81	6.23	3.80	8.42	1.18
'R138 net'PYR.p -->PEP.p	-2.00	-29.84	19.89	12.69	-10.31	-16.80	-6.16	2.71
'R154 net'3PG.c -->3PG.p	-14.97	-46.09	32.73	20.11	-16.95	-34.27	1.86	9.22
'R155 net'DHAP.c -->DHAP.p	-1.58	-52.50	55.34	27.51	-5.74	-23.87	8.59	8.28
'R156 net'PEP.c -->PEP.p	6.75	-24.60	31.03	14.19	14.88	4.15	24.39	5.16
'R157 net'PYR.c -->PYR.p	10.64	-16.45	34.85	13.09	-3.63	-11.70	9.51	5.41
'R158 net'G6P.c -->G6P.p	-3.68	-11.45	3.81	3.89	-1.47	-5.22	1.44	1.70

10.9.4. Influence of data coverage on flux identifiability

Table 10-17 (next page): Fluxes calculated with different subsets of the entire data set available. No GCMS: all measurements from GC-MS analysis were omitted. noAA: all measurements of amino acids were omitted for the flux calculation. noPoolSizes: Pool size quantification was omitted. All fluxes were determined by ^{13}C-INST-MFA and are expressed in mmol $(100 \text{ mmol } CO_2)^{-1}$. The boundaries of the 95 % confidence intervals and the thereof derived standard deviation were determined through parameter continuation.

Reaction number	noGCMS				noAA				noPoolSizes			
	val	lb	ub	std	val	lb	ub	std	val	lb	ub	std
'R1'	91.57	91.41	92.08	0.17	91.67	91.41	92.58	0.30	91.53	91.41	91.82	0.10
'R2'	5.84	5.30	6.16	0.22	5.71	5.09	6.24	0.29	5.89	5.43	6.24	0.21
'R3'	85.70	85.26	86.30	0.27	85.94	85.18	86.89	0.43	85.64	85.19	86.07	0.23
'R4'	8.43	7.92	8.59	0.17	8.33	7.42	8.59	0.30	8.47	8.18	8.59	0.10
'R5'	0.29	0.00	0.63	0.16	0.58	0.00	1.17	0.30	0.21	0.00	0.42	0.11
'R6'	2.68	2.68	2.68	0.00	2.68	2.68	2.68	0.00	2.68	2.68	2.68	0.00
'R7 net'	-3.47	-3.52	-3.45	0.02	-3.70	-3.90	-3.45	0.11	-3.46	-3.49	-3.45	0.01
'R8 net'	4.06	4.04	4.16	0.03	4.21	4.04	4.49	0.11	4.05	4.04	4.11	0.02
'R9 net'	-4.06	-4.16	-4.04	0.03	-4.21	-4.49	-4.04	0.11	-4.05	-4.11	-4.04	0.02
'R10 net'	73.12	72.84	73.68	0.22	73.70	73.05	74.33	0.33	73.23	72.87	73.92	0.27
'R11 net'	-182.55	-183.89	-181.78	0.54	-182.70	-185.57	-181.97	0.92	-182.93	-184.49	-181.85	0.67
'R12 net'	17.93	17.59	18.11	0.13	17.96	17.55	18.11	0.14	17.97	17.52	18.12	0.15
'R13 net'	17.93	17.59	18.11	0.13	17.96	17.55	18.11	0.14	17.97	17.52	18.12	0.15
'R14 net'	-7.70	-8.33	-7.34	0.25	-14.30	-19.18	-7.34	3.02	-13.06	-19.13	-7.34	3.01
'R15'	2.17	2.17	2.17	0.00	2.17	2.17	2.17	0.00	2.17	2.17	2.17	0.00
'R16'	0.06	0.00	0.21	0.05	0.24	0.00	0.45	0.11	0.05	0.00	0.17	0.04
'R17'	0.06	0.00	0.21	0.05	0.24	0.00	0.45	0.11	0.05	0.00	0.17	0.04
'R18 net'	35.98	35.82	36.25	0.11	36.01	35.80	36.57	0.20	35.98	35.83	36.35	0.13
'R19 net'	-35.98	-36.25	-35.82	0.11	-36.01	-36.57	-35.80	0.20	-35.98	-36.35	-35.83	0.13
'R20 net'	33.14	32.98	33.42	0.11	33.17	32.97	33.73	0.20	33.14	32.99	33.51	0.13
'R21'	100.86	100.46	101.72	0.32	100.93	100.62	102.71	0.53	101.08	100.50	102.20	0.44
'R22'	100.00	100.00	100.00	0.00	100.00	100.00	100.00	0.00	100.00	100.00	100.00	0.00
'R23'	69.12	68.80	69.67	0.22	69.18	68.77	70.30	0.39	69.11	68.82	69.85	0.26
'R24'	0.86	0.46	1.72	0.32	0.93	0.62	2.71	0.53	1.08	0.50	2.20	0.44
'R25'	0.86	0.46	1.72	0.32	0.93	0.62	2.71	0.53	1.08	0.50	2.20	0.44
'R26'	0.55	0.00	1.24	0.32	0.14	0.00	1.83	0.47	0.65	0.00	1.78	0.46
'R27'	0.55	0.00	1.24	0.32	0.14	0.00	1.83	0.47	0.65	0.00	1.78	0.46
'R28'	0.47	0.47	0.53	0.02	0.47	0.47	0.54	0.02	0.47	0.47	0.51	0.01

	C1	C2	C3	C4	C5	C6	C7	C8	C9	C10	C11	C12
R29 net'	3.01	6.68	-5.10	-1.26	3.04	6.78	-5.14	-1.69	0.25	-4.14	-5.10	-4.90
'R30 net'	0.90	0.82	-2.73	-0.09	10.23	0.77	-39.34	-12.19	0.30	0.50	-0.69	0.05
'R31 net'	8.27	-1.83	-34.26	-5.92	9.13	-1.92	-37.73	-16.91	0.33	-2.15	-3.44	-2.70
'R32 net'	9.12	33.04	-2.70	6.41	9.07	33.03	-2.53	9.30	0.28	-1.22	-2.34	-1.84
'R33 net'	0.00	0.00	0.00	0.00	0.00	0.00	0.00	0.00	0.00	0.00	0.00	0.00
'R34 net'	3.01	2.71	-9.07	-1.13	3.04	2.75	-9.17	-0.71	0.25	2.70	1.74	2.50
'R35 net'	10.29	0.81	-39.51	-10.00	10.23	0.77	-39.34	-12.19	0.30	0.50	-0.69	0.05
'R36'	0.15	2.71	2.12	2.56	0.14	2.71	2.14	2.55	0.13	2.70	2.19	2.53
'R37'	0.15	2.71	2.12	2.56	0.14	2.71	2.14	2.55	0.13	2.70	2.19	2.53
'R38 net'	0.15	2.70	2.10	2.55	0.14	2.69	2.13	2.54	0.13	2.69	2.18	2.51
'R39'	0.15	0.60	0.00	0.45	0.14	0.59	0.03	0.44	0.13	0.59	0.07	0.41
'R40'	0.15	0.59	0.00	0.44	0.14	0.59	0.02	0.43	0.13	0.58	0.07	0.41
'R41'	0.15	0.97	0.38	0.82	0.14	0.97	0.40	0.81	0.13	0.96	0.45	0.79
'R42'	3.01	11.79	0.00	5.72	3.02	11.84	0.00	6.96	0.25	0.99	0.00	0.36
'R43'	8.27	32.43	0.00	4.09	9.13	35.89	0.09	15.08	0.33	1.61	0.32	0.86
'R44'	0.00	0.10	0.10	0.10	0.00	0.10	0.10	0.10	0.00	0.10	0.10	0.10
'R45'	0.00	0.00	0.00	0.00	0.00	0.00	0.00	0.00	0.00	0.00	0.00	0.00
'R46'	0.30	7.49	6.29	6.82	0.51	7.93	5.92	6.69	0.42	8.07	6.42	7.13
'R47'	0.30	5.60	4.40	4.93	0.51	6.05	4.03	4.80	0.42	6.19	4.54	5.24
'R48'	0.00	0.06	0.06	0.06	0.00	0.06	0.06	0.06	0.00	0.06	0.06	0.06
'R49'	0.01	0.51	0.47	0.48	0.02	0.54	0.47	0.49	0.02	0.53	0.47	0.49
'R50'	0.01	0.51	0.47	0.47	0.02	0.54	0.47	0.47	0.02	0.53	0.47	0.47
'R51'	0.00	0.03	0.03	0.03	0.00	0.03	0.03	0.03	0.00	0.03	0.03	0.03
'R52'	0.00	0.15	0.15	0.15	0.00	0.15	0.15	0.15	0.00	0.15	0.15	0.15
'R53'	0.00	0.00	0.00	0.00	0.00	0.00	0.00	0.00	0.00	0.00	0.00	0.00
'R54'	0.15	0.62	0.03	0.15	0.14	0.60	0.07	0.23	0.09	0.36	0.05	0.09
'R55'	0.00	0.05	0.05	0.05	0.00	0.05	0.05	0.05	0.00	0.05	0.05	0.05
'R56'	0.19	0.97	0.22	0.54	0.14	0.65	0.11	0.32	0.17	0.91	0.23	0.43
'R57'	0.05	0.54	0.33	0.47	0.09	0.93	0.58	0.69	0.11	0.82	0.39	0.57
'R58'	0.28	1.17	0.07	0.51	0.29	1.33	0.19	0.38	0.24	1.00	0.07	0.31
'R59'	0.28	1.17	0.07	0.51	0.29	1.33	0.19	0.38	0.24	1.00	0.07	0.31

'R60'	0.58	0.58	0.58	0.00	0.58	0.58	0.58	0.00	0.58	0.58	0.58	0.00
'R61'	0.41	0.41	0.41	0.00	0.41	0.41	0.41	0.00	0.41	0.41	0.41	0.00
'R62'	0.55	0.55	0.55	0.00	0.55	0.55	0.55	0.00	0.55	0.55	0.55	0.00
'R63'	0.30	0.00	0.62	0.16	0.17	0.00	0.62	0.16	0.24	0.00	0.62	0.16
'R64'	0.32	0.00	0.62	0.16	0.45	0.00	0.62	0.16	0.39	0.00	0.62	0.16
'R65'	0.13	0.13	0.13	0.00	0.13	0.13	0.13	0.00	0.13	0.13	0.13	0.00
'R66'	0.34	0.34	0.34	0.00	0.34	0.34	0.34	0.00	0.34	0.34	0.34	0.00
'R67'	0.38	0.38	0.38	0.00	0.38	0.38	0.38	0.00	0.38	0.38	0.38	0.00
'R68'	0.33	0.33	0.33	0.00	0.33	0.33	0.33	0.00	0.33	0.33	0.33	0.00
'R69'	0.35	0.35	0.35	0.00	0.35	0.35	0.35	0.00	0.35	0.35	0.35	0.00
'R70'	0.27	0.27	0.27	0.00	0.27	0.27	0.27	0.00	0.27	0.27	0.27	0.00
'R71'	0.09	0.09	0.09	0.00	0.09	0.09	0.09	0.00	0.09	0.09	0.09	0.00
'R72'	0.77	0.77	0.77	0.00	0.77	0.77	0.77	0.00	0.77	0.77	0.77	0.00
'R73'	2.06	2.06	2.06	0.00	2.06	2.06	2.06	0.00	2.06	2.06	2.06	0.00
'R74'	6.64	6.34	6.97	0.16	6.51	6.34	6.97	0.16	6.58	6.34	6.97	0.16
'R75'	0.00	0.00	0.00	0.00	0.00	0.00	0.00	0.00	0.00	0.00	0.00	0.00

10.10. Measured pool sizes

As direct pool sizes could potentially increase flux identifiability, the concentrations of several intracellular metabolites was determined by LC-MS/MS or GC-MS.

Table 10-18: Metabolic pool sizes of several intracellular metabolites in the wild type rice seedling, as well as in DMSO and imazapyr treated plants.

μmol (g DW)$^{-1}$	wild type	DMSO	imazapyr	method
PEP	0.92 ± 0.06	1.06 ± 0.17	0.93 ± 0.09	LCMS
AKG	1.83 ± 0.28	2.13 ± 0.17	1.72 ± 0.42	GCMS
P5P	1.80 ± 0.27	1.69 ± 0.19	1.53 ± 0.10	LCMS
S7P	4.21 ± 0.16	3.51 ± 0.46	2.24 ± 0.13	LCMS
SUCC	1.23 ± 0.06	1.48 ± 0.26	0.77 ± 0.04	GCMS
2PG	1.88 ± 0.14	1.56 ± 0.29	1.62 ± 0.09	LCMS
3PG	42.02 ± 2.88	36.58 ± 2.31	35.17 ± 5.71	LCMS
FUM	0.43 ± 0.02	0.42 ± 0.01	0.37 ± 0.02	GCMS
E4P	0.20 ± 0.06	0.17 ± 0.02	0.07 ± 0.04	LCMS
RBP	0.15 ± 0.03	0.21 ± 0.05	0.17 ± 0.03	LCMS
alanine	10.70 ± 0.82	8.12 ± 0.35	15.76 ± 0.12	GCMS
valine	1.11 ± 0.03	0.85 ± 0.03	0.32 ± 0.02	GCMS
leucine	0.41 ± 0.03	0.35 ± 0.01	< LOD	GCMS
isoleucine	0.31 ± 0.03	0.27 ± 0.02	0.33 ± 0.00	GCMS
glycine	3.60 ± 0.28	2.84 ± 0.22	3.90 ± 0.07	GCMS
proline	0.69 ± 0.03	0.47 ± 0.03	1.55 ± 0.02	GCMS
serine	7.27 ± 0.18	5.91 ± 0.38	7.93 ± 0.04	GCMS
threonine	3.08 ± 0.18	2.41 ± 0.17	3.78 ± 0.03	GCMS
aspartate/ asparagine	18.43 ± 0.39	17.24 ± 0.75	23.58 ± 0.56	GCMS
methionine	0.33 ± 0.02	0.25 ± 0.01	0.35 ± 0.00	GCMS

glutamate/ glutamine	114.55 ± 7.33	88.08 ± 4.62	145.07 ± 3.85	GCMS
inositol	4.65 ± 0.27	4.74 ± 0.17	5.33 ± 0.07	GCMS
tyrosine	0.62 ± 0.08	0.58 ± 0.02	0.81 ± 0.03	GCMS
lysine	0.42 ± 0.07	0.45 ± 0.02	0.53 ± 0.01	GCMS
glucose	13.07 ± 2.17	11.86 ± 1.73	11.28 ± 0.51	GCMS
fructose	19.02 ± 1.08	19.02 ± 2.74	17.70 ± 0.92	GCMS
phenylalanine	0.41 ± 0.00	0.35 ± 0.01	0.58 ± 0.01	GCMS

10.11. Futile cycles imposed by energy dissipation

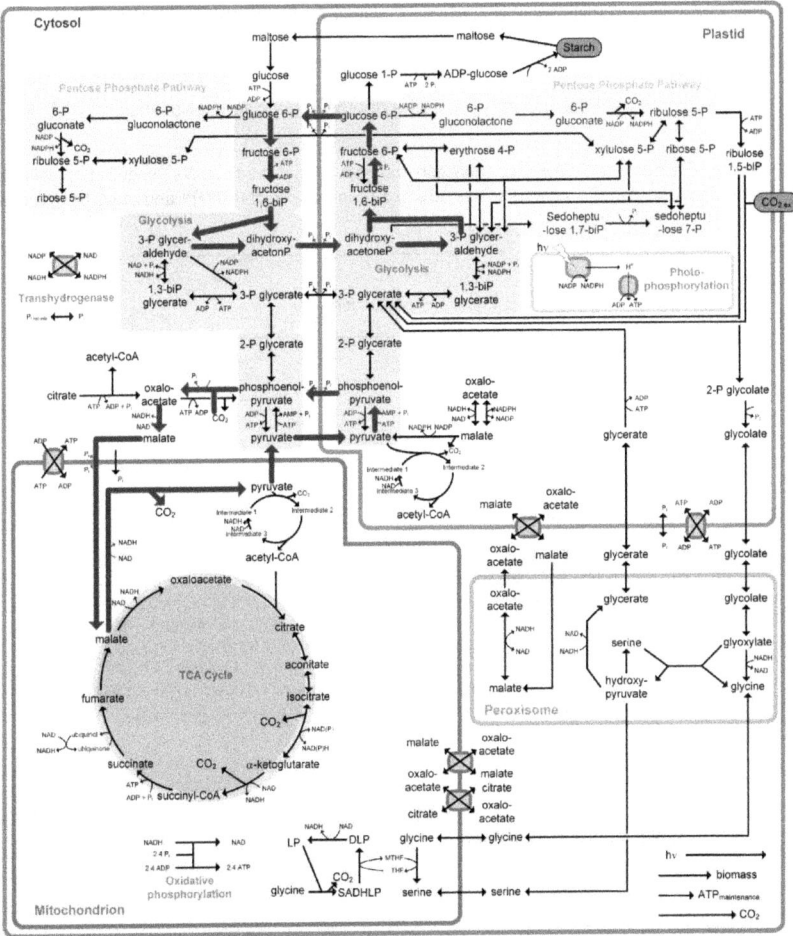

Figure 10-2: Two exemplified futile cycles in the proposed metabolic network causing energetic inefficiency

The futile cycle across the mitochondrial membrane consumes one plastidial ATP per cycle, whereas the futile cycle involving the two copies of the EMP pathway requires one cytosolic ATP per cycle.

10.12. Stoichiometric reaction network of *O. sativa* seedling shoots

The previously developed network model of *O. sativa* seedling shoots was now adjusted for application in EFM analysis. For this purpose, the metabolic network model was extended with redox and energy metabolites, as well as with the photosynthetic light reactions.

Table 10-19: Finalized stoichiometric reaction network for *Oryza sativa* seedling metabolism. Stoichiometric factors of the biomass equation are expressed as mmol $(gDW)^{-1}$ and atom transitions are represented between brackets.

Reaction number	Reaction
	CO_2 metabolism
1	-> CO2EX[p]
2	CO2[p] -> CO2EX[p]
3	CO2[p] ->
	Biomass Synthesis (untreated)
4	0.1181 PYR[cp] + 0.1979 AKG[m] + 0.4437 P5P[p] + 0.1299 OAA[c] + 0.0433 3PG[cp] + 0.0218 F6P[cp] + 0.3704 G6P[cp] + 0.1083 GAP[cp] + 0.0055 CIT[m] + 0.0012 SUCC[m] + 0.0342 MAL[m] + 0.0206 FRC[cp] + 0.7041 GLC[cp] + 0.1765 sucrose[cp] + 0.0377 starch[cp] + 0.2171 alanine[c] + 0.1417 arginine[c] + 0.2327 aspar[c] + 0.0173 cysteine[p] + 0.3163 glutamate[m] + 0.2341 glycine[p] + 0.0483 histidine[p] + 0.1022 isoleucine[p] + 0.2047 leucine[p] + 0.1229 lysine[p] + 0.0337 methionine[p] + 0.7690 phenylalanine[p] + 0.1282 proline[c] + 0.0377 serine[p] + 0.1297 threonine[p] + 0.2891 tyrosine[p] + 0.1548 valine[p] + 0.0001 tryptophane[p] + 0.0098 C16[p] + 0.0573 C18[p] + 0.001075 C20[p] + 1.9445 NADPH[cp] + 5.3753 ATP[cp] -> biomass + 1.9445 NADP[cp] + 5.3753 ADP[cp]
5	biomass -->
	Embden-Meyerhof-Parnas pathway
6	G6P[cp] <-> F6P[cp]
7	F6P[cp] + ATP[cp] --> FBP[cp] + ADP[cp]
8	FBP[cp] --> F6P[cp]
9	FBP[cp] <-> DHAP[cp] + GAP[cp]
10	GAP[cp] <-> DHAP[cp]
11	GAP[cp] + NADP[cp] + ADP[cp] <==> ATP[cp] + NADPH[cp] + 3PG[cp
12	3PG[cp] <-> 2PG[cp]
13	2PG[cp] <-> PEP[cp]
14	PEP[cp] + ADP[cp] --> PYR[cp] + ATP[cp]
15	PYR[cp] + ATP[cp] --> PEP[cp] + AMP[cp]
16	AMP[cp] + ATP[cp] --> 2 ADP[cp]
17	PYR[cp] + E2Pr-lip[cp] --> E2Pr-acet-lip[cp] + CO2[p]
18	E2Pr-acet-lip[cp] --> ACCOA[p] + E2Pr-2hyd-lip[cp]

19	E2Pr-2hyd-lip[cp] + NAD[cp] --> E2Pr-lip[cp] + NADH[cp]

Oxidative pentose phosphate pathway

20	G6P[cp] + NADP[cp] --> NADPH[cp] + 6PG[cp]
21	6PG[cp] + NADP[cp] --> NADPH[cp] + P5P[p] + CO2[p]

Non-oxidative pentose phosphate pathway

22	GAP[cp] + S7P[p] <-> E4P[p] + F6P[p]
23	E4P[p] + P5P[p] <-> F6P[p] + GAP[cp]
24	GAP[cp] + S7P[p] <-> 2 P5P[p]

Calvin-Benson-Bassham cycle

25	P5P[p] + ATP[cp] --> RBP[p] + ADP[cp]
26	CO2EX[p] + RBP[p] -> 2 3PG[cp]
27	DHAP[cp] + E4P[p] -> S7P[p]

Photorespiration

28	RBP[p] -> GLYCO[pg] + 3PG[cp]
29	GLYCO[pg] -> GLYOX[pg]
30	2 GLYOX[pg] + 2 NADH[pg] + serine[p] -> glycine[p] + GLYCER[pg] + 2 NAD[pg] + CO2[p]
31	GLYCER[pg] + ATP[cp] --> 3PG[cp] + ADP[cp]
32	MAL[pg] + NAD[pg] <==> OAA[pg] + NADH[pg]

Amphibolic reactions

33	OAA[c] + ATP[cp] --> CO2[p] + PEP[cp] + ADP[cp]
34	PEP[cp] + CO2[p] --> OAA[c]
35	MAL[p] + NADP[cp] <-> NADPH[cp] + PYR[cp] + CO2[p]

Transporters

36	MAL[m] <-> MAL[c]
37	PYR[m] <-> PYR[cp]
38	OAA[m] + CIT[c] <-> OAA[c] + CIT[m]
39	MAL[m] + OAA[c] <-> MAL[c] + OAA[m]
40	MAL[c] <-> MAL[p]
41	ATP[m] + ADP[cp] --> ADP[m] + ATP[cp]
42	MAL[c] --> MAL[pg]
43	OAA[pg] --> OAA[c]

Mitochondrial metabolism

44	PYR[m] + E2Pr-lip[m] --> E2Pr-acet-lip[m] + CO2[p]
45	E2Pr-acet-lip[m] --> ACCOA[m] + E2Pr-2hyd-lip[m]
46	E2Pr-2hyd-lip[m] + NAD[m] --> E2Pr-lip[m] + NADH[m]
47	OAA[m] + ACCOA[m] -> CIT[m]

48	CIT[m] + NAD[m] <==> AKG[m] + CO2[p] + NADH[m]
49	AKG[m] + NAD[m] --> SUCCCOA[m] + CO2[p] + NADH[m]
50	SUCCCOA[m] + ADP[m] --> SUCC[m] + ATP[m]
51	SUCC[m] + UQN[m] --> FUM[m] + UQL[m]
52	UQL[m] + NAD[m] <==> UQN[m] + NADH[m]
53	FUM[m] -> MAL[m]
54	MAL[m] + NAD[m] <--> OAA[m] + NADH[m]
55	MAL[m] + NAD[m] --> CO2[p] + NADH[m] + PYR[m]

Starch synthesis

56	G6P[cp] + ATP[cp] -> starch[p] + ADP[cp]

Sugar metabolism

57	G6P[cp] -> GLC[cp]
58	GLC[cp] + ATP[cp] -> G6P[cp] + ADP[cp]
59	F6P[cp] -> FRC[cp]
60	G6P[cp] + F6P[cp] + ATP[cp] -> sucrose[cp] + ADP[cp]
61	sucrose[cp] -> sucrose.r
62	sucrose.r -->

Fatty acid metabolism

63	8 ACCOA[p] + 14 NADPH[cp] + 8 ATP[cp] -> C16[p] + 14 NADP[cp] + 8 ADP[cp]
64	9 ACCOA[p] + 16 NADPH[cp] + 9 ATP[cp] -> C18[p] + 16 NADP[cp] + 9 ADP[cp]
65	10 ACCOA[p] + 18 NADPH[cp] + 10 ATP[cp] -> C20[p] + 18 NADP[cp] + 10 ADP[cp]

Amino acid metabolism

66	3PG[cp] + 2 ATP[cp] -> serine[p] + 2 ADP[cp]
67	3PG[cp] + 4 NADPH[cp] + 4 ATP[cp] -> cysteine[p] + 4 ADP[cp] + 4 NADP[cp]
68	3PG[cp] + NADPH[cp] + 3 ATP[cp] -> 3 ADP[cp] + glycine[p] + CO2[p] + NADP[cp]
69	GLYOX[pg] + NADPH[cp] -> glycine[p] + NADP[cp]
70	glycine[p] + 3 NADH[m] -> MTHF[p] + CO2[p] + 3 NAD[m]
71	glycine[p] + MTHF[p] -> serine[p]
72	PYR[cp] + 2 ATP[cp] + glutamate[m] -> 2 ADP[cp] + alanine[c] + AKG[m]
73	2 PYR[cp] + 2 ATP[cp] + NADPH[cp] + glutamate[m] -> valine[p] + AKG[m] + CO2[p] + 2 ADP[cp] + NADP[cp]
74	2 PYR[cp] + ACCOA[p] + glutamate[m] + 2 ATP[cp] -> 2 ADP[cp] + leucine[p] + 2 CO2[p] + AKG[m]
75	OAA[c] + 5 ATP[cp] + glutamate[m] -> aspar[c] + AKG[m] + 5 ADP[cp]
76	OAA[c] + NADPH[cp] + 2 ATP[cp] -> aspar[c] + 2 ADP[cp] + NADP[cp]
77	P5P[p] + CO2[p] + glutamate[m] + 2 NADP[cp] + 6 ATP[cp] -> histidine[p] + AKG[m] + 6 ADP[cp] + 2 NADPH[cp]
78	AKG[m] + 2 NADPH[cp] + NADH[m] + 3 ATP[cp] + ATP[m] -> proline[p] + 2 NADP[cp] + NAD[m] + 3 ADP[cp] + ADP[m]
79	glutamate[m] + OAA[c] + CO2[p] + NADPH[cp] + NADH[m] + 4 ATP[cp] + 3 ATP[m] -> arginine[c] + FUM[m] + NADP[cp] + NAD[m] + 4 ADP[cp] + 3 ADP[m]
80	OAA[c] + PYR[cp] + glutamate[m] + 3 NADPH[cp] + 3 ATP[cp] + ATP[m] -> lysine[p] + 3 NADP[cp] + 3 ADP[cp] + ADP[m] + CO2[p] + AKG[m]
81	OAA[c] + 3 NADPH[cp] + 4 ATP[cp] -> threonine[p] + 3 NADP[cp] + 4 ADP[cp]
82	OAA[c] + PYR[cp] + glutamate[m] + 4 NADPH[cp] + 4 ATP[cp] -> isoleucine[p] + CO2[p] + 4

	NADP[cp] + 4 ADP[cp] + AKG[m]
83	OAA[c] + 3PG[cp] + CO2[p] + 11 NADPH[cp] + 9 ATP[cp] -> methionine[p] + 11 NADP[cp] + 9 ADP[cp] + PYR[cp]
84	E4P[p] + 2 PEP[cp] + glutamate[m] + 3 ATP[cp] -> tyrosine[p] + 3 ADP[cp] + CO2[p] + AKG[m]
85	E4P[p] + 2 PEP[cp] + glutamate[m] + NADPH[cp] + 3 ATP[cp] -> phenylalanine[p] + 3 ADP[cp] + NADP[cp] + CO2[p] + AKG[m]
86	AKG[m] + NADH[m] -> glutamate[m] + NAD[m]
87	E4P[p] + 2 PEP[cp] + P5P[p] + 3PG[cp] + glutamate[m] + 2 NADPH[cp] + 6 ATP[cp] -> CO2[p] + 6 ADP[cp] + 2 NADP[cp] + tryptophane[p] + PYR[cp] + GAP[cp] + AKG[m]

Energy metabolism

88	2 Hv[p] + PQN[p] --> PQL[p] + 2 H_nc[p]
89	PQL[p] + 2 PC_o[p] <==> PQN[p] + 2 PC_r[p] + 2 H_nc[p]
90	Hv[p] + PC_r[p] + FE_o[p] --> PC_o[p] + FE_r[p]
91	2 FE_r[p] + NADP[cp] --> 2 FE_o[p] + NADPH[cp] + 2 H_nc[p]
92	Hv[p] --> 2 H_c[p]
93	2 H_c[p] + 12 H_nc[p] + 3 ADP[cp] --> 3 ATP[cp]
94	ATP[cp] --> ADP[cp] + ATP_maint[cel]
95	(2.4) ADP[m] + NADH[m] --> NAD[m] + (2.4) ATP[m]
96	--> Hv[p]
97	ATP_maint[cel] -->
98	NADP[cp] + NADH[cp] <==> NADPH[cp] + NAD[cp]'

www.ingramcontent.com/pod-product-compliance
Lightning Source LLC
Chambersburg PA
CBHW060254220326
41598CB00027B/4093